Adventures in Managing Water

Adventures in Managing Water

Real-World Engineering Experiences

Edited by Daniel (Pete) Loucks and Laurel Saito

Sponsored by

Environmental and Water Resources Institute of
the American Society of Civil Engineers

PUBLISHED BY THE AMERICAN SOCIETY OF CIVIL ENGINEERS

Library of Congress Cataloging-in-Publication Data

Names: Loucks, Pete, author.
Title: Adventures in managing water : real-world engineering experiences / edited by Pete Loucks.
Description: Reston, Virginia : American Society of Civil Engineers, [2019] | Includes bibliographical references and index.
Identifiers: LCCN 2019011530| ISBN 9780784415337 (softcover : alk. paper) | ISBN 9780784482278 (pdf)
Subjects: LCSH: Watershed management—Case studies. | Water-supply—Case studies.
Classification: LCC TC413 .A38 2019 | DDC 628.1—dc23
LC record available at https://lccn.loc.gov/2019011530

Published by American Society of Civil Engineers
1801 Alexander Bell Drive
Reston, Virginia 20191-4382
www.asce.org/bookstore | ascelibrary.org

Illustrations by Angela Liu and Alicia Wang.
Interior design and composition by Robert Kern, TIPS Technical Publishing, Inc.

Errata: Errata, if any, can be found at https://doi.org/10.1061/9780784415337.

ISBN 978-0-7844-1533-7 (print)
ISBN 978-0-7844-8227-8 (PDF)

Manufactured in the United States of America.

25 24 23 22 21 20 19 1 2 3 4 5

Contents

Foreword

We who inhabit this earth are here because of the earth's natural resources. Water is one of them, and we all know that without water, life and many of the goods and services we enjoy would not exist. But having access to water in the right amounts and qualities to meet our needs is not always possible without the infrastructure we design and operate and the institutions we create to manage it. Managing the water nature gives us can be a challenge due to the way water is linked to our social structures, economies, ecological systems, and landscapes. It requires and involves individuals trained in multiple disciplines. Engineering is one of them. In this book, some of us who have devoted much of our professional careers, whether just beginning or nearly ending, to water resources planning and management, share some of our experiences addressing issues related to water. The scale of these issues range from local to international.

We have written these essays in hopes that they will be of interest to and even entertain you. Our stories are written especially for those beginning their professional training. There are lots of lessons and moments of enlightenment in these essays. They deal with a wide variety of water projects and experiences encountered by engineers, planners, economists, and social scientists. Some of our attempts to address them were successful, some were not. But all of them involved an adventure—dealing with people directly impacted by what we were doing.

All of us in this profession have a number of stories that, if told, would be both inspirational and educational, as well as fun to hear or read. We who are involved in water resources systems planning and management are indeed fortunate to have each other as colleagues. We admit that much of what we know and do, and get invited to do, results from our association with each other. We who are in this "club" invite our young student readers to join. We need you to help us meet society's needs for clean, reliable water supplies, deal with floods and droughts, and work toward reducing the uncertainties and conflicts among people that characterize our work. Perhaps these vignettes will give you a hint about how managing water and addressing related issues can change peoples' lives as well as their natural environments.

Introduction

We suspect you may be wondering just what this book is about. We considered two groups of potential readers. The first group is the one we set out writing for—young, eager, bright, adventurous, ambitious students who are, or will soon be, beginning their college or university education and who have to decide, eventually, just what discipline or subject to study. All of us who have contributed to the stories contained in this book were in this group, either a few or many years ago. For whatever reason or circumstance, we all ended up working on issues dealing with the management of water. Our purpose in telling some of our adventures involving water is to convince all of you why you should consider preparing yourself to face the challenges, surprises, and satisfactions we have had addressing important water management issues. Providing adequate and reliable and affordable and clean water supplies is necessary for any well-functioning society.

Water is a big deal for all of us on this planet. Can you think of anything that doesn't require water to make or exist? The management of the water we use, and the design and operation of the infrastructure and institutions we create to do it, at a community or global scale, can challenge all our technical and social skills. Trying to satisfy everyone's needs for water, when both future supplies and demands are unknown and changing, is never boring. Our stories in this book attest to that observation. If you choose this profession, you will be working on complex interdisciplinary problems that affect society and our environment. It will force you to continue to grow intellectually, both as an engineer or scientist, but most importantly, as a person.

The second group of readers of this book may indeed be all of us who have had our own adventures dealing with water and are interested in the experiences of others dealing with different water management issues in different places in the world. What strikes many of us who were mostly trained as engineers is that in so many cases the social challenges of managing water often dominate the technical ones. Building institutions and creating water management policies, whether in Iraq following a war, or coping with droughts, floods, or pollution issues in places like Australia, Brazil, Chile, China, Germany, Hungary, Iran, Jordan, Libya, Laos, Morocco, Poland, Portugal, Russia, South Africa, Tunisia, and Uzbekistan—to mention a few of the countries where our stories take place—is a social process. We learn the same is true whether fixing water pipes within a prison or establishing consistent and effective policies for reaching sustainability goals among international organizations.

Reading these stories will make us appreciate again how what we do can make a difference in peoples' lives. How a few gave their lives saving so many others after the accident at Chernobyl. How managing Lake Baikal in Russia, the largest freshwater lake by volume in the world, presented many of the same challenges involved in establishing policies for regulating the Great Lakes in North America, or Lake Balaton in Hungary, or even Lake Winnipeg and the Red River in Canada. And we learn how managing flows and their allocations on the Mississippi River in the US, the Mekong River in Southeast Asia, the Vistula River in Poland, the Danube River in Europe, and even the Great Man-Made River in Libya is about much more than obeying the physical laws of hydrology and hydraulics. Likewise, managing water quality is much more than applying the principles of biology, chemistry, and environmental engineering. We also reflect on how the decisions we made early in our careers led, eventually, to the experiences we can now write about. All of these stories show that our profession is as much about people and politics as it is about water science and technology. Whatever group you are in, we believe and hope you will find reading these brief essays both educational and entertaining.

This idea of preparing short essays addressing some of our adventures is admittedly a rather different kind of writing exercise for many of us. The criteria we set for ourselves were that each story had to be short and free of jargon (that any professional literature is full of) and references (that every scholarly paper must have)—clearly everything that would be grounds for rejection by any respected peer-reviewed professional journal. To everyone who has contributed to this book, our thanks for reinforcing our feeling that what we do is not only valued by society but is also an adventure.

Part I

Adventures in Africa

Water Planning in Egypt: Reusing the Nile

Leon Hermans

There are several countries where the importance of water is directly visible and strongly intertwined with a country's history and culture. Egypt is such a place. Since ancient times, the Nile River has played a critical role in the economic and cultural life of Egyptian society. Current satellite images show that population and economic activity still concentrate along a relatively small strip on the borders of the Nile River and into the Nile Delta. Although the Nile is a majestic and impressive river, its flows are limited. Hence, the freshwater resources of Egypt require careful and thoughtful management.

Around the turn of the millennium, the government of Egypt was preparing a new national water resources plan to ensure sustainable management of its water resources. One promising approach being explored was the reuse of drainage water from agriculture. Water reuse is not merely a technical issue but a social one as well. At that time, decision-making regarding reuse involved multiple governmental agencies, including the Ministry of Water Resources and Irrigation, the Ministry of Agriculture, the Ministry of Health and Population, the National Organization of Potable Water Supply and Sewage Disposal, and the Egyptian Environmental Affairs Agency. Their interests and ideas about drainage water reuse often differed.

As you can imagine, this multi-stakeholder complexity posed additional challenges to the water resources planning team in Cairo. I was to support the team with an analysis that would capture important decision-making features around agricultural drainage water reuse. Developing models to support these decision-making processes is notoriously difficult and sensitive. We decided to start first with capturing and structuring the understanding and information available within the planning team. For this, I used an actor-modeling approach based on game theory. This had a strong internal validity and logic. In addition to model building, I carried out a data collection program using structured interviews. A first version of the model was presented for discussion in an interactive workshop with the team. The participants played out parts of the decision-making process captured by the model. This turned out to be a most surprising experience.

The model was based on structured interviews with the planning team members. However, during the workshop, the participating team members made choices and arguments that completely countered what they had shared earlier during the interviews. What happened here?

Before the workshop, when asked about the organizations involved in drainage water reuse and the options these organizations had to influence decision-making around the issue, the planning team's experts sketched the decision-making process as they understood it to be. However, they mainly referred to the procedure as they thought it ought to be, in theory.

During the interactive workshop, they represented decisions and power of organizations as they had experienced them in the past. For instance, the Ministry of Agriculture had very little official involvement in drainage reuse; officially, its decision making was mainly about ensuring the quality of the water being reused or released to the environment. However, in the workshop, this ministry seemed to be almost at the center of decision-making. As participants explained, "In the end, the Minister of Agriculture is the most influential Minister in the Cabinet. He will have to agree, or the decisions will not be accepted. Everyone knows this." This made good sense, given the economic and societal importance of agriculture in the country. However, this had not come up during the interviews. The planning team members hadn't realized that this was important, or that I didn't know about this common fact, nor had I been able to ask for it; it is hard to ask for something if you don't know what that something is.

In the end, the session helped me improve the actor model in important ways. It turned out to be a session for model *development,* not a session to *present* and explain the model results, with only some final tweaking of the model afterward. You might think that this experience is unique to the use of actor models or other social research connected to water systems engineering. I don't think so. Of course, social, economic, and hydrological—or hydraulic—models all differ, but the lesson I took from this experience applies to most, if not all, models that we use as analysts and planners to support our clients.

Reflecting on the workshop in Cairo, I have come to appreciate that our models are the instruments that help us to make sense of the information we have and sometimes to identify information we should have. Modeling enables us to probe for a deeper understanding of what we should know or do. Sense-making and probing are best done with a group of people. Critically discussing models and modeling results is at least as important as developing models and producing results. We are often producing something similar to a composite sketch of a wanted person's face. Based on secondary information, we try to reconstruct a fitting picture of reality. Imagine a police composite sketch that is not checked back with some of the key witnesses. You wouldn't want to rely on this, no matter how skillful and diligent the sketch artist has been. Reconstructing parts of a water system—physical, decision making, or otherwise—is not that different.

Ever since that experience, I have come to value direct discussions and interactions with stakeholders, experts, and decision makers about not just model outputs but also modeling data and assumptions. Such discussions, with both experts and their clients, give me new insights and fresh perspectives. From them I gain knowledge. Model building and use can guide and inform these discussions, as this experience illustrates. And this is exactly why I model.

The Vision Kid

Leif Lillehammer

Vision Framework

Vision Kid

I followed in the footsteps of my father, who instilled in me his interest in biology and environmental sciences. I have spent most of my professional career as a freshwater ecologist and water resources practitioner. Not a surprise, probably, since I have visited most, if not all, small, medium, and large rivers in Norway, starting from my childhood. We were often looking for stoneflies (Plecoptera), one of my father's specific research interests, on which he became a renowned expert.

At my university, I began studying astronomy, motivated by the popular TV series called *Cosmos.* I loved *Cosmos,* and Carl Sagan, its writer and narrator, was my big hero in the early/mid-1980s. But after a while, I bounced back to finish my studies in the field of freshwater ecology. After graduating in June 1991, I got a position at the University of Oslo, Zoological Museum, as a scientist doing freshwater and fisheries ecology research. I worked there for about five years.

In early 1996, I left the Zoological Museum to join a consultancy company called Statkraft Engineering, but I continued, from time to time, doing research at my former workplace with my earlier colleagues.

By the mid-1990s, some early versions of the World Wide Web had emerged and were available at various universities. There were some active scientific groups using the internet to communicate with each other at that time. As part of a Norwegian Research Council project on hydropower peaking, I posted a question to one of these groups about peaking effects on various fish species. An answer came from Mark Bain, an ecologist at Cornell University. This led to a frequent exchange of ideas on a variety of subjects. Over time, we became friends and research colleagues. I invited Mark over for a workshop in Oslo in 1997, and Mark invited me over to Cornell three times between 1998 and 2002. Mark also worked and collaborated with Pete Loucks. Mark and Pete influenced me vastly during this period, with regard to water resources management. I brought their ideas into practice in consultancy projects I have participated in since that time. These projects have included addressing water resources management issues in Asia, Africa, and Central and South America.

Just before and after the millennium shift, I worked on water resources management issues in Central America, including developing a basin plan for the rUlua and Chamelecon rivers in Honduras after Hurricane Mitch. In late 2001, my geographic focus shifted to East Africa. By the end of the year, my company, together with Cornell University and some other research institutes, won a contract to develop *The Vision and Strategy Framework for Sustainable Development of Lake Victoria Basin.* This was a massive two-year exercise covering the countries of Uganda, Kenya, and Tanzania, with close to 20 regional and international experts on the team. We worked with stakeholders from local, district, and national levels up to the regional (East African Community) level in helping them define their "desired future for the lake basin, and how to get there," hence the term *The Vision and Strategy Framework.* Thanks to our client, the East African Community, we also got help from other stakeholders throughout the study. Close to 13,000 people were consulted in the process.

The most rewarding consultations were those at the local level. For example, at the Wakawaka fish village in Uganda, I talked to those who fished for their food and income. Their visions and strategies were debated and harmonized at each step, from the local up to the regional level. In the autumn of 2003, the regional vision statement was accepted by the three countries at the East African Community headquarters in Arusha, Tanzania. The regional vision was further adopted by the Council of Ministers in January 2004. *The Vision and Strategy Framework for Sustainable Development of Lake Victoria Basin* has been a point of departure for other initiatives of the Lake Victoria Basin Commission ever since. This project would end up being a turning point in my life.

Parallel to this story comes the emergence of the Vision Kid. Originally, I was to coordinate the international expert team and the country team of Kenya. By pure chance, the latter was shifted to Uganda in early 2002, where I lived through the summer of 2003. This shift changed my life completely. I met my future wife, and we moved to a house in the outskirts of Kampala. We had our wedding, a mix of African and Norwegian styles, in February 2003. Team members, the Royal Norwegian Embassy, and representatives from the Toro Kingdom (western Uganda) participated in our wedding, along with some 300 other guests—a small wedding in Ugandan terms. We moved back to Norway just before our daughter Amalie was born. At the regional workshop in Arusha in the autumn of 2003, everybody knew about Amalie, and the East Africans cordially named my daughter "the Vision Kid" because of the project. I have frequently returned to this region over the years, and now, in 2018, I am back doing work for the Lake Victoria Basin Commission, while my Vision Kid is soon turning 15.

Wuthering Heights—Tales from the Kingdom in the Sky

Leif Lillehammer

"My work has taken me to some interesting places."

I spent the years 2006 and 2007 living in the Kingdom in the Sky. I led part of a larger project there during these two years, the Lesotho Water

Sector Improvement Project. Lesotho is a land of breathtaking beauty and utterly nice and friendly people. I had around eight trips to Lesotho in this period, some with my family in the first half of 2006.

Flying into Maseru from Johannesburg—especially in the European summertime, which is wintertime in Lesotho—is a breathtaking experience since you see the magnificent snowcapped Drakensberg mountains. Furthermore, touring the Lesotho highlands with its wuthering heights with peaks close to 3,500 m, and its lush green landscapes during the wet season, sometimes feels like the Garden of Eden. Maseru itself is in the lowlands, at around 1,500 m, and you can't get much lower than that in Lesotho. Thus, Lesotho is known as the country with the highest lowest elevation above mean sea level (amsl) in the world. It's also a constitutional monarchy, so its inhabitants term it "the Kingdom in the Sky." During the project period, I had multiple trips to the highlands, and in the grandest tour, we drove a four-wheel drive across most of it. During this trip, we also passed the Sani Pass at 2,876 m amsl and stayed over at the lodge with the same name. Over the lodge bar was a badge: Highest Pub in Africa.

My work with the Lesotho Water Sector Improvement Project covered multiple aspects of developing the water sector, including offering institutional and regulatory advice. Our main task was to develop a water policy and a Lesotho Integrated Water Resources Management (IWRM) strategy. For the latter, concepts from the textbook by Pete Loucks and Eelco van Beek, and others, on water resources planning and management were used. Early in 2007, I became the policy advisor to the Commissioner of Water. That lasted until the project ended later the same year.

I made many friends in this wonderful country. I was back in 2011 on a short visit for another project and in 2017 for yet another one. The last part of the Lesotho Highlands Water Project (LHWP), one of the most massive and long-lasting water projects in the world, is still ongoing. These trips always involve having a jolly good time with my Lesotho friends. The joys of consulting!

Digging a Well in Africa

Hugo A. Loáiciga

"Why the well? We have lake water!"

In early 2011, I was invited by a nongovernmental organization (NGO) to serve as a technical adviser for a team of about ten volunteers who would assist with the installation of a groundwater well in a rural, equatorial region of Africa. The team made two journeys to the country, each lasting about four weeks. The first journey involved making logistical arrangements and collecting data pertinent to the installation and operation of the well. The second journey was devoted to creating the well and installing the pipes, storage tanks, and other infrastructure that would allow water users to extract groundwater with a gasoline powered generator and pump water to various discharge points, whose distances from the well varied from a few to about 500 m.

The well installation was done by a local (African) company. The basic hydrogeological study assessing the well siting, depth, design, production rate, and specific capacity had been completed by a local geologist. The well would extend to a depth of 120 m. The groundwater quality was suitable for human use, except for a relatively high concentration of fluoride. Some type of remedy would have to be found.

I was told by the NGO that the well installation had been approved by the pertinent African authorities. There was also support from a board of directors led by a tribal chief. The NGO had obtained funding for the well project, and the community to be served by the well had between 4,000 and 6,000 inhabitants. Most of them lived in small huts in an area of about 4 km^2. There was a small central village containing a few crumbling buildings with small shops, a military garrison, and a clinic built by another American NGO.

Except for a few government-paid teachers and functionaries, most of the inhabitants were impoverished. Many had irregular work at best;

some engaged in rain-fed farming, while others fished. This population has been historically affected by AIDS, which is still carried by many of its members. The clinic built by an American NGO provides basic medical and hospice services to AIDS victims. I saw many AIDS victims and many orphan children while walking in the area. An elementary orphan school educates and provides basic services to many of these orphans. There was no electricity service, public potable water, or sewage system. There were a few schools, however, and most people I met were literate. The closest food market was about 16 km away. The living conditions for most people were poor, yet many appeared at peace with their situation. Our team had some social celebrations and interactions with the local inhabitants that were unforgettably pleasant. This proved to me that despite poverty, it is possible for humans to enjoy uplifting moments.

A notable geographic feature of this community is the large, natural, freshwater lake bordering it. This has been the historic water source for people who fetch water from the lake with buckets carried by donkeys and vehicles. People's main means of locomotion are walking, bicycles, and donkeys.

When first considering my participation in the well project, I asked why the NGO wanted to install a groundwater well 1.5 km from a large freshwater lake that historically served everyone for their basic water needs. I never received a satisfactory response. The lake has a high fluoride concentration, just as the groundwater does, but, unlike the groundwater, it was also subjected to biological contamination. Nevertheless, the lake has been the source of water for humans in this region since time immemorial. Moreover, the well's water yield would be very small compared with the 10–20 liters per person of household use in this area. Nevertheless, community leaders welcomed outside money invested in community projects, and the NGO was engaging in other beneficial charitable activities.

The trip from my hometown in the United States to this village in Africa was an odyssey. It started with a thirteen-hour flight to the capital of a Middle Eastern country, followed by a five-hour flight to the capital of the African country. There was an overnight stay at a lodge in the capital. The next day started with a one-hour flight from the country's capital to the provincial capital. From there I took a three-hour boat ride to a small port where a pickup truck took us to the village. The boat ride was spectacular. The boat had space for passengers and for domesticated animals, such as cattle, all managed in what I can best describe as controlled chaos. The village leaders had arranged the rental of a small house with bunk beds where the team members crowded in during their stay. Protection against mosquito and fly biting was a constant struggle and necessary precaution. The team paid for a couple of local cooks and helpers to take care of meal preparation, some house cleaning, and carrying out a few indispensable chores, such as bringing bottled water for our use. There were children following us wherever we went.

During the groundbreaking ceremony, just before the drilling rig began operating, a young man claiming to be the son of the man who owned the lot where the well would be installed became disruptive. He claimed his late father never granted permission for the well to be installed in his lot. The young man was belligerent and waving a machete. He was eventually subdued and whisked away by tribal leaders and relatives. Obviously, the nuances of well permitting and securing land access for drilling had not been totally resolved beforehand. There were also minor acts of sabotage whereby some trenches were refilled and water pipes punctured, and plumbing supplies stolen. And the drilling contractor and electrician interrupted work for hours or several days, claiming he had to go tend to other matters.

Despite these incidents and delays, the well water system was completed during our team's second journey. There was a closing ceremony, and training was given by team members to locals who would operate the well and maintain its associated water system.

I learned immensely from my two journeys to the African country. Primarily, I witnessed a very different lifestyle, and I developed a much better appreciation of how good life is in my home country. One has to experience the contrasts to understand them. In the months following the second journey, I worked hard to secure funding to improve the school for orphan children. The tribal leader confessed to me during a social event that the main community priority was expansion of the school's capacity and improvement of sanitation there. The well construction was not a community priority.

In the months since my return from the second journey, I have learned the village has had trouble paying for the gasoline needed to power the generator for the well. Community residents had to share the cost of buying and transporting the gasoline, because there was no communal or governmental funding to pay for energy provision. The more groundwater being used, the larger the drop in the water level inside the well. This in turn increased the energy, and hence cost, required to fill the water tanks. I estimate the gasoline cost was in the tens of dollars daily; thus, in a place where daily per capita income may average two dollars, the provision of gasoline was evidently a burden. Most residents could take their donkey to the lake and fetch water without any cash outlay.

This project illustrates the fact that even if a project is paid for by some outside agency, it is not always cost-free to the recipients who are to supposedly benefit from it. I realize the NGO opted to pay for the well because this type of small project is within its domain of expertise. Here, there was failure on the part of the NGO and local leaders to think through the well project and explore other higher-priority projects that would best benefit the community. As a systems analyst, I should have been more critical of what we were doing and more proactive in suggesting to the NGO other ways to invest their good will and money in this village. But how can one know this when first being asked by any aid agency to help? Even though this experience was unforgettable and educational from a personal viewpoint, I have to conclude it is a good example of misplaced technology.

Water for Food in North Africa

Pete Loucks

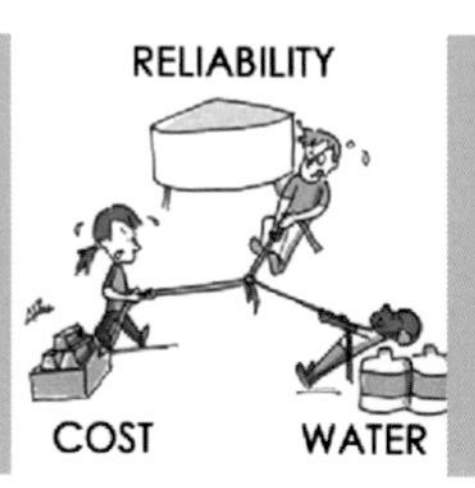

During the 1973 Arab–Israeli War, Arab members of the Organization of Petroleum Exporting Countries (OPEC) imposed an oil and gas embargo against the United States and other countries friendly to Israel. This contributed to an upward spiral in oil and gas prices. The price of oil per barrel first doubled, then quadrupled, imposing skyrocketing costs on consumers and threatening the stability of the impacted economies.

Algeria was, and remains, a member of OPEC. As the flow of money resulting from exports of Algeria's oil and gas increased, their Ministère de l'Industrie et de la Promotion des Investissements began to think about the possibility of a trade war, where countries that Algeria depends on to supply food could raise their food prices in response to higher gas and oil prices. The question being asked in Algeria was how could it use the money they were making to become more agriculturally self-sufficient should such a trade war, or embargo, occur in the future. The solution the ministry came up with was to spend that surplus of money on water resources infrastructure that would "capture every drop of water that would otherwise flow into the Mediterranean Sea and transport it over the Atlas Mountains to irrigate and produce food from the northern parts of the Sahara Desert—regardless of cost." Soil studies indicated that crops could grow in the desert if water was available to irrigate them.

A water resources engineering firm in Madrid, Spain, was given the contract to plan the development of that infrastructure. I was asked by that firm to do the analyses needed to identify the siting, design, and operating policies for all the reservoirs, pumping stations, wastewater treatment plants, and pipes such infrastructure would entail throughout the country. This project kept a group of my graduate students busy for about three years. Our offices were full of computer paper outputs containing the information we were asked to write up and deliver. I traveled to Madrid once per month, getting there on a Friday to deliver our results based on our current contract and then pick up the next month's contract and instructions in time to leave on Sunday. The firm in Madrid translated all our results into French and eventually published all of our analyses in fancy gold trimmed hardcover books—again "regardless of cost!"

Our job as systems analysts was to identify the infrastructure Algeria needed to maximize the amount of water that could be provided for irrigating crops at various sites in the Sahara Desert. As we thought about that, it occurred to us that a range of amounts could be provided, and all would be correct depending on their reliability. The response we got when we asked what reliability Algeria wanted, was, "That's your job; you are the experts." Not knowing how to pick the "best" reliability, we gave them a range of tradeoffs among the amounts of water and their reliabilities. One can promise more water if its reliability is less. In addition, we felt that even though Algeria seemed not too concerned about the cost, its officials might judge the cost of delivering that last drop of water rather excessive. Thus, the tradeoffs we gave them were among three opposing criteria: cost, amount, and reliability of irrigation water.

To actually grind out all possible combinations of cost, water delivered, and its reliability for numerous irrigation sites in the basin would have taken us more time than we had, so we devised a means of interacting with some people in the ministry to help us locate the region of solutions (tradeoffs among all three criteria) of most interest to them. This was before the days of email between my university and Algeria. Hence, this interactive process took weeks at each step. In one case, officials seemed to like a solution that cost more but provided less water and less reliability than a range of other possible solutions. We asked why they thought it best, and they told us of another objective they had in that particular part of their country, namely regional development. We learned to expect this, not only in this project but in almost every other one I've been associated with. The beauty of this systems approach to planning is that as stakeholders and clients learn more about how their systems work, or could work, they will have new ideas about how they want it to work. We analysts must adapt as we inform. If our clients and stakeholders are listening and reacting to what we produce, I judge that to be a measure of success.

Was this project successful? A month after the Madrid firm delivered its multiple gold-plated reports, the president of Algeria died. The government changed. I'm told the new government judged the previous government incompetent, ignored reports prepared by or for various ministries, including ours, and gave a similar contract to another engineering firm. That firm contacted me and asked for the results of our part of the study. We sent it to them, but I've heard nothing since, and I still have never been to Algeria.

Water from the Desert in Libya

Pete Loucks

"We were looking for oil, but we found something more valuable -- water."

Have you ever wanted to visit Libya? I had the opportunity to go to Libya at the time the "Brotherly Leader and Guide of the Revolution" Muammar Gaddafi was building his Great Man-Made River. This "river" is a network of groundwater wells, pumps, and pipes that supply water found deep under the sand of the Sahara Desert. This so-called fossil water is similar to other nonrenewable resources in that it is not being replenished. But there is plenty of it. Libya is using that water mainly to irrigate crops, so as to become more agriculturally self-sufficient.

The *New York Times* once headlined on page 1 that this network of 4 m diameter pipes extending from one side of the country to the other was really being built to transport trains, troops, tanks, and trucks to the borders of Libya for possible military operations without being seen by satellites. In reality, the Great Man-Made River Project is an amazing water delivery system that has changed lives of Libyans all across the country. The cost has exceeded 25 billion US dollars, and all of it has been paid by Libyans.

The Great Man-Made River project is divided into five phases. After the first two phases of the project were built, the Libyan government asked UNESCO to evaluate the plans for further development and compare their costs to other options for obtaining water. As part of a team to complete that task, two of us went to Libya to observe the progress made to date and to obtain the economic and engineering data we needed to determine the most cost-effective ways of meeting various water demand targets at specified sites and times. This got us a tour into the pipe network itself (before it began transporting water) and into the offices of the foreign firm doing all the design work. But we also needed data from various ministries, and typically those ministries did not want to give us their valuable data. The only option I could think of to get to know and gain the trust of some of those working in those ministries located in Benghazi was to join the Benghazi squash club where many of their young employees played squash. It worked. They won most of our games, but they gave us our needed data. We completed our analyses, which showed that the cheapest way to meet the water demand targets was to continue developing the Great Man-Made River. To come to that conclusion, our analyses had to define the capacities, locations, and staging of various components of the further phases of the system of pipes, pumps, and reservoirs, and their sensitivity to future costs and interest rate uncertainties. We did not consider the possibility of conflict that now exists in Libya. We can only hope the current conflicts will end soon so the Libyans can enjoy the water, and hence the food, the Great Man-Made River can provide.

A Water Management Story from South Africa

Ronnie Mckenzie

"Yeah, we can't build this."

In the early 1980s, South Africa and Lesotho developed a water transfer scheme called the Lesotho Highlands Water Project that was designed to transfer water from the Lesotho Highlands to Vaal Dam via two parallel 5 m diameter transfer tunnels approximately 85 km long. The transfer tunnels were designed to transfer 70 m^3/s through the Maluti mountains into the upper reaches of the Ash River that, in turn, flowed to the main Vaal Dam from where the water would be used to supply the industrial powerhouse of Africa. At the time the scheme was developed, South Africa was under considerable pressure from the rest of the world through sanctions aimed at removing the Apartheid government. One of the few "friends" South Africa had at the time was the tiny mountain kingdom of Lesotho, which is surrounded by South Africa. Agreeing on what was to become one of the largest water transfer projects in the world between the two countries was very important to South Africa, not only because it needed the additional water supply but also to show the rest of the world that South Africa was still open for business and had some friends left.

Against this backdrop of political and financial considerations, the Lesotho Highlands Water Project was finally conceptualized in the early 1980s. The project was designed to be implemented in five phases involving five reasonably large dams between 120 m and 185 m in height, as well as two 80 km long transfer tunnels, each approximately 5.5 m in diameter. This scheme was the water engineer's dream of what could be achieved through cooperation between two countries who depended on each other for their future prosperity. South Africa needed the water and had money to pay for the transfer scheme, whereas Lesotho had the water but needed funding to support its mainly rural population. Both countries had very good reasons to support the new water project, which was hailed internationally as one of the most significant engineering projects of the twentieth century.

My role in the project started in 1983 when I was encouraged to take up employment in South Africa to help develop computer systems models capable of analyzing complicated water supply systems. The rapid advancements in computing power had created great opportunities in the fields of hydrological modeling and the subsequent system modeling. Working together with specialists from Canada, United States, United Kingdom, and, of course, South Africa, I was part of an impressive team of world-recognized experts whose sole aim was to create a new model that could be used to analyze and manage South Africa's highly complicated water resource infrastructure. After four years of continuous effort, models were developed that were capable of analyzing large and very complex water resource systems using a combination of network modeling together with a very sophisticated stochastic streamflow generator.

In 1987, I was given the opportunity to test out the new models on what is still considered to be one of the largest and most complicated water resource systems in the world. The Orange-Senqu water resource assessment was commissioned by the South African government, and in 1991, we delivered the draft report of our findings. As a relatively young and inexperienced water engineer, the opportunity to analyze the Orange-Senqu river basin was challenging to say the least. During my studies in Scotland, I was working on the Stanford Watershed Model from the United States and testing the snow modeling routines on two catchments of 50 km^2 and

250 km^2 respectively. Moving up to the Orange-Senqu river basin with a catchment area of over 1,000,000 km^2 required some change in mind-set. We once calculated that if we spent the same effort on the Orange-Senqu basin as we had spent on the Scottish basins, it would have taken us almost 10,000 years to complete the work. Using the new models and cutting a few corners, we did manage to trim the analysis down to approximately four years, which was bearable and within budget.

The South African government had expected the new models to confirm the viability of the massive new water project, and the report was considered a necessary formality in order to persuade the World Bank and others to assist in funding the new dams and transfer tunnels. Unfortunately, the results were not as expected and showed that the Orange River would basically run out of water along the lower reaches of the river somewhere between Phase 2 and 3 of the Lesotho Highlands Water Project. This result was very significant and highly sensitive. On receipt of the report, the top 12 water managers within the South African government convened a "feedback session," which was similar to a trial where I (and two colleagues) had to sit on one side of a very large table answering questions on how we were proposing canceling the largest water transfer scheme in the world despite the many studies by international specialists who all said it was viable. This potentially career-limiting meeting extended for several hours and resulted in an embargo on the report until such a time as the South African government could obtain a second or third opinion—hopefully ones that contradicted our findings. After six months of careful investigation, the government's own Department of Hydrology produced a report that was virtually identical in its conclusions to that from our team. They confirmed that the Lesotho Highlands Water Project could not proceed beyond Phase 2 without creating major water problems along the lower Orange River and into Namibia. Our first major study using the new system analysis models had resulted in canceling (at least in part) what would have been one of the largest water transfer projects in the world—hardly something to add to a CV!

Almost 30 years later, the Lesotho Highlands Water Project has been constructed as far as Phase 1 with a subsequent Phase 2 currently under construction and due for completion around 2025. The second set of transfer tunnels and additional dams are not being considered at the moment, and the scheme has been curtailed to a two-phase scheme. Curtailing the original five-phase scheme remains one of the most significant studies I have completed in almost 40 years of water engineering. It may not be as satisfying as building a new dam somewhere, but it is nonetheless great to see that people can still take up rafting tours of the lower Orange River and have sufficient water in the river to navigate the rapids.

The South African government accepted the results and now insists that any new development be analyzed thoroughly using the very same system models that were developed back in the 1980s by various specialists from Canada (Rick Allen and Oscar Sigvaldason), the United States (Prof. Pete Loucks, Prof. Jerry Stedinger, and Prof. Neil Grigg), the United Kingdom (Peter Adamson and Prof. O'Connel), and South Africa (Prof. Pegram, Steffan van Biljon, Johan van Rooyen, Thinus Basson, and Pieter van Rooyen).

The Ultimate Water Manager

H. P. Nachtnebel

"Water managers sure are celebrated around here."

In early 2000, a group of us engineers was invited to Morocco to propose principles and guidelines for establishing river basin authorities in Tunisia. The European Union water framework directive was already implemented in Europe, and this document provided the basis for our proposal. We began our work by having several meetings with experts and politicians at government ministries designed to come to an agreement on objectives and principles. We also identified the need to develop requirements for developing a nation-wide database, the data it should contain, and a tool box that would facilitate its use.

Another of our activities during our visit was to explore Morocco's river basins and to meet with local water managers and users to learn about their expectations with respect to water management. We had several meetings at various villages to learn about traditional irrigation practices. In one of the villages in the headwater area, we were invited to the house of a regional traditional water manager. Of course, males and females were separated into different lunch rooms.

After having had an excellent lunch accompanied by mint tea, the water manager explained to us his principles, duties, and daily actions. We learned that the water manager holds a socially prestigious position within the local community. Hence, many are interested to be the elected water manager and to execute this job. The local water community, which consists of all water consumers, elects the water manager. He must be male and have some land-related water demands (e.g., irrigation). He has full power to allocate water resources among the users, and based on his expertise he could allocate the full amount of available water to a single user over a short term if he wishes. In case anyone is unsatisfied with his or her water allocation, for example because of water shortage, he or she can approach the water community and demand a revised allocation of water. If the community agrees, a new water manager would be elected. Of course, this would bring disgrace to the former water manager and, it turns out, also to his family.

In a private discussion with the water manager, I expressed my opinion that the best strategy for a water manager would be to satisfy the water demands in an equitable way to all water users to avoid conflict among the individual users. He responded that he followed this approach in principle, but sometimes in drought periods a particular farmer might need a disproportionately high water supply because his crops were in a very final stage and any reduction of supply would result in major economic losses. Thus, the water manager has the duty to inspect the status of the crops frequently and to meet with those who might suffer from water shortage to negotiate an acceptable allocation of water.

I argued that it seemed a good strategy would be to satisfy most of the users, even disregarding an individual demand, because the probability of being replaced by someone else would be low. He agreed but then answered that this strategy would create mistrust within the village community, which would be disadvantageous for all water users. So something like general consent in the applied strategy should be an overall goal. A "disproportionate" allocation of water—in other words, a favored allocation—should be explained and discussed among all the users to obtain their agreement, hopefully. Finally, he told me that he had already been reelected water manager several times and thus thought that his strategy had worked well, even though he had had to favor some users on occasion when their harvests would have failed otherwise.

After lunch was over, one female member of our group returned from the other dining room. She mentioned that she had enjoyed the lunch and the conversation among the local women. Apparently, from their dining room they could listen to our conversations in the men's dining room. She told us later that the Moroccan women were always critical of and mocked the statements of their husbands. Also, as they, being women, were responsible for serving the food, they had decided to keep the best parts of the meal for themselves while serving the rest to the male group.

I like the Moroccan water management approach, which is similar to the scissors-paper-stone strategy. Water managers are elected by the water user group, their position and function is very prestigious, they have unlimited power with respect to water allocation, and they can be forced out of their position at any time.

From Working on the Railroad to Hydrologic Modeling—A Life of Lessons

Geoff Pegram

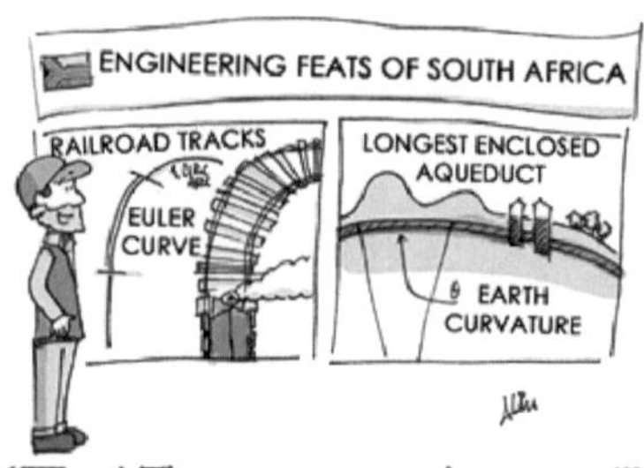

"Wow! Those are some nice curves!"

My early career as a professional engineer was full of surprises, and like all new graduates trying to do things right with little experience, I was on a steep learning curve. I was first employed by Rhodesia Railways in 1962 to honor the scholarship they granted me to obtain my bachelor's degree in civil engineering. My first railway job was in Bulawayo involving track renewal between Bulawayo and Victoria Falls. My supervisor had previously worked as a road engineer, so his solution for replacement, against all protestations from his practically minded and experienced railway subordinates, was to remove all the old stone ballast and muck from the railroad foundation. His plan was to recompress the track foundation with vibratory rollers as used on highways. This decision turned out to be a disaster because the foundation was fine sand with very little cohesion and collapsed under load—the "muck" was actually valuable as it provided cohesion and integrity of the formation. We nearly lost a couple of trains before he was sacked. Railway foundations are different from roads in that they are vibrated to compaction by the heavy trains over time. I had never been taught the difference between these in the "structures" and "design" courses as an undergraduate. Railway engineers are a different breed and learn from their seniors' and forebears' collective experience on the job. So the first important lesson I learned was to listen to competent, experienced, and practical people, even if they were not "degreed."

My next lesson was learned when I was transferred to Broken Hill (now Kabwe) in Zambia. I was put "in charge" of track renewal between Broken Hill and Ndola (about 164 km) under a very experienced professional railway civil engineer called Jack Carlisle. Incidentally, later, I had the pleasure to teach his son hydraulics at the University of Natal. Jack introduced me to the powerful management tool of "time and motion studies." It was vital that I mastered this tool, because the new track-renewal program had been stepped up a pace. I was "technically" (at the age of 24) in charge of a very large organization, and the person directly responsible to me was a Permanent Way inspector called Giovani Fossataro (Johnny). He was old enough to be my father, so he deserved to be listened to. Subordinate to him were 13 plate-layers, mostly Italian and highly respectful of Johnny, who had all remained in Rhodesia after the Second World War—they were very experienced in rail track work. They in turn supervised about 300 track laborers in cohorts of different tasks who were responsible, within a window of two hours between trains, to remove half a mile of track (comprising 40 ft long rail tracks, fish-plates, and wooden sleepers) and replace them with fresh sleepers and 120 ft rail tracks. These 44 rail sections then had to be screwed down atop the fish-plates onto the new sleepers—all in two hours. In addition, fresh stone ballast that had been off-loaded from ballast trains a few days earlier had to be lashed onto the track and quickly vibrated by hand-operated Plasser machines to compact the ballast to make the line safe, all before the next train could come through. I'm glad to say that we met that challenge, most of the time.

One of the technical problems, almost unique to railways, that intrigued me was the way that the curves were designed to change the track's direction. In those days, the technique adopted by the railways surveyor was to lay out the transition following a cubic curve. I analyzed this curvature from the point of view of rate of change of radius to see how smooth the transition was for passengers and cargo and found that there would always be a slight jolt in the transition. My idea was that the most comfortable transition would come from using a Euler curve. I laid out my methodology in a small pamphlet that earned me the Rhodesia Railways prize for the best innovation of the year. Today, this is the standard practice in the railway industry worldwide.

After fulfilling my obligation to the railway company, my next job was back in South Africa working on ORT, the Orange-Fish tunnel (5.4 m in diameter and 84 km long), which was the longest continuous enclosed aqueduct in the southern hemisphere and, possibly still, the second-longest interbasin water supply tunnel in the world. It was designed to provide water to a parched area in South Africa called the Transkei and the city of East London. The tunnel was so long it had to be aligned on a constant-radius vertical curve, following the Earth's curvature, or else it would never have been possible to empty it. In addition, the theodolites used to set out the targets for drill and blast had to be set with centrifuges spun up with high speed to be able to accurately determine geographical direction at the bottom of the five 500 m vertical shafts. These multiple access shafts allowed the construction of several sections of the tunnel to proceed simultaneously.

The third lesson I learned at the ORT involved filling water-supply pipelines (40 cm diameter) to test their integrity and water-tightness. We used a pump to suck the water (that looked like milky coffee) from the Orange River to fill the pipe for pressure testing over an elevation change of 230 m. There were two surprises for me, the novice engineer: (1) the pump, being

2 m above the river level, suffered from cavitation, so it could not draw water, and (2) after we set the pump on a raft in the river, the pressure at the pump fluctuated because of the elasticity of the cement pipe, which was about 3 km long. This meant we had to monitor maximum pressure to keep it within tolerance. Live and learn!

Once the construction of the ORT tunnel was complete, I left to be a lecturer at my alma mater, the University of Natal. Sometime after I had established myself in the lecturing position, the senior consulting engineer on ORT called me. He asked me how he could determine the hydraulic roughness of a new tunnel he was boring without using a lining. This tunnel is a 37 km delivery tunnel from the Muela Reservoir in the Lesotho Highlands to the outfall at the Ash River from where water flows to the Vaal Dam in South Africa. Much of the tunnel length was in fine unfractured sandstone, which needed no dressing. The question the engineer asked me was, "How hydraulically rough is a bored unlined tunnel?" There were no available guidelines that he could find. My master's degree student, Mark Pennington (who got a distinction for this work), spent a lot of time in the tunnel taking laser scans of the wall over about 10 km of tunnel. His data were used successfully in the tunnel's design. As an indication of its value, last year I had a request from Europe for a copy of Mark's thesis.

As an associate professor, I was quite happy and comfortable doing my lecturing, supervising postgraduates, going on sabbaticals, doing research, speaking, and publishing. Then, out of the blue in 1985, I was approached by Thinus Basson (one of the authors in this set of stories) to spend part of my six-month sabbatical with his team putting together the stochastic hydrological model for the water resources system of South Africa. At the time, South Africa was experiencing its worst drought on record. Pete Loucks recommended me to Dr. Basson to design the software to model multisite stochastic streamflow gauges and rain gauges over South Africa. The upshot was that I put together the suite of computer programs for analyzing multisite streamflow and rainfall time series, first in the Vaal River Basin and then over all of South Africa. That software is still used to this day, 33 years later. I shamefully have to admit that while Pete was visiting us in Pretoria, he always beat me at squash racquets! All is forgiven; Pete was responsible for asking me to write this, my story.

More academically, about that time in 1985, I was asked by Dr. George Green, then assistant director of the South Africa Water Research Commission, to enter into hydrometeorology research contracts with them involving (1) merging of radar and satellite estimates of daily rainfall, (2) developing a "High Resolution Space-Time" rainfall modeling approach, (3) producing "Daily Rainfall Mapping over South Africa," and more. Over 34 years, I completed 21 projects for the WRC, ending in 2018 with the report on modeling soil moisture and evapotranspiration over the SADC Region.* I love my work and amazingly, even get paid for it!

Navigating in the Estuaries of Mozambique

Hubert Savenije

"Measuring salinity is quite the adventure!"

As a recently graduated hydraulic engineer, my first job was in Mozambique, where I worked from 1978 to 1985. There, I developed my initial theories on salt intrusion in estuaries. Many years later, this research resulted in my Ph.D. and my online book, *Salinity and Tides in Alluvial Estuaries* (salinityandtides.com). Yet doing all this wasn't in my mind at the time. When I arrived in Mozambique, I was asked to study the cause of the salinity intrusion in the estuaries of the Limpopo and the Incomati, where farmers during the dry season suffered from poor quality water. The first thing I decided to do was measure the salinity along the estuary. I wanted to find out what made the salinity travel all the way upstream. To obtain consistent data, we considered it essential to measure salinity at the moment of high-water slack. This occurs about half an hour after

* Southern African Development Community (SADC) is a regional economic community comprising 15 member states: Angola, Botswana, Democratic Republic of Congo, Lesotho, Madagascar, Malawi, Mauritius, Mozambique, Namibia, Seychelles, South Africa, Swaziland, Tanzania, Zambia, and Zimbabwe.

high water. The tidal wave travels with the speed of a speedboat, so I needed one. I was allowed to import a sturdy 65 HP Boston Whaler, which I named *Afra* after my recently born daughter. During the following years, I surveyed four different Mozambican estuaries with *Afra*. The data I collected proved invaluable; I still use them to test and improve my theories.

However, all good things come to an end. In 1985, I left Mozambique, and I had to leave *Afra* behind. I started working for a consulting firm in Delft, the Netherlands. Fortunately, a couple of years later, an opportunity arose. A Delft University graduate student, Marieke, wanted to do research on salt intrusion in estuaries, and preferably in Africa. Some of my consulting work involved returning to Mozambique, so I took that opportunity to involve her in taking more salinity measurements on the Incomati estuary. I found *Afra* in a garage, where it had been for a few years. It looked good and its engine still ran. Early the next morning, we set out to the mouth of the Incomati estuary, and at high tide, we started taking our measurements following the tidal wave. At full throttle, *Afra* broke through the water beautifully, but then things went wrong. After an hour, the engine suddenly stopped and, with a shock, the boat came to a standstill. Apparently, the cooling system was not working properly; the engine had overheated and then stalled. There we were, floating without a paddle, in the Incomati. Using our hands as paddles, we brought *Afra* to the shore.

The Incomati estuary channel meanders quite strongly. We had travelled perhaps 20 km, but this was only about 5 km as the crow flies. We decided we could easily walk back to our car in about an hour through the sand dunes of the meander. But what was the direction? We knew the time, so with the position of the sun and a simple map (no smartphones with GPS at that time) we set our course. The sun was now high in the sky, it was hot, and we had hardly any water with us. After half an hour of walking, I got suspicious. The landmarks that I was expecting to see did not appear. And then suddenly I realized we were walking in the wrong direction. With the northern sky in my head, I had forgotten that in Mozambique the sun is in the north, not the south as it is when observed at that time of year from the northern hemisphere. In my head I had swapped north and south—a mistake that seemed inexcusable for someone who had lived in Mozambique for six years.

After having realized this mistake, we set course for our car to meet our colleagues. They were happy to see us again, and the feelings were mutual. We survived, and who knows—we may have even benefited from the extra exercise walking in the sand dunes of Incomati. Our measurements were incomplete, but those that we had were put to good use later on.

A few years later, I joined a university, and Marieke became my first Ph.D. student.

A água sabe o caminho (The Water Knows the Way)

Hubert Savenije

"Ahh!! The water knew the way!"

"The water knows the way." This is what an old man in Mozambique told me when, as a young engineer, I visited his village to help them restore a dike breach. He suggested it was no use to repair the dike because the water would breach it again at that very location.

It was in the early 1980s, during a flood that threatened to inundate the floodplain of the Limpopo river in Mozambique, when the old man spoke these words to me. We were inspecting the dike that protected his village from flooding. As a young engineer and as a Dutchman, I had high confidence in human-engineered flood protection works. Dikes were meant to protect people and dike failure did not fit within my professional vocabulary.

"The water knows the way. It will continue to hit the dike until it breaks. And then it will rejoin its old river course." The old man, who had seen many floods, hit the air with his fists to underline his words. I smiled at the obvious ignorance of the old man while he smiled at mine. To me, he was just an illiterate old man who had no idea about river hydraulics or the complexities of river engineering. To him, I was just another young know-it-all, riding a white Land rover, who had not yet lived long enough to know the ways of the water and to understand its unpredictable nature: sometimes gentle—bringing life, coolness, and cleanliness to the rural community; and sometimes, at occasions like these, demonstrating sheer violence and the power of death and destruction.

To me, water doesn't know anything; it is merely a collection of molecules moving under the influence of gravity, and it certainly has no will of its own. A person who attributed a spirit to water was clearly uneducated, if not plain stupid. How wrong I was. At that time, I didn't see that it was me who was ignorant and that I should have been grateful that the old man was sharing his wisdom with me while knowing I wouldn't understand.

During that flood, the dike stood firm. Although it was threatened, we managed to secure the dike, strengthening it with sand bags and maintaining a dike watch until the water subsided. We had proven the old man wrong.

Twenty years later, an unprecedented flood hit Mozambique, and the village of the old man was completely submerged by 2 m of water. The dike had broken at exactly the same location as the old man had indicated. After the flood subsided, there was a mild stream of water following the old river course through the gap that the dike failure had left. The old man had been right, not in a literal sense of course, but his metaphor had been disturbingly precise.

Lately, I have thought a lot about that old man, and I feel ashamed, not about how I acted, because during our brief encounter I had remained respectful and friendly, but about how I had felt. I had the sheer arrogance of assuming I knew better just because I benefited from a better education and access to information. Now that I have become an old man myself, I can only hope that some of the young engineers that I have taught will have the wisdom to listen to the words of old men and remember them long enough to appreciate their significance.

Developing an Acceptance of Systems Methods for Water Resources Planning and Management in South Africa

Oskar Sigvaldason and Thinus Basson

Water resources planning in South Africa is an exercise in dealing with extremes. Droughts of record can be immediately followed by floods of record. This gives rise to the need for large impoundments, control structures, and interbasin diversions for retaining runoff during wet periods, and for meeting water-based demands during dry periods.

The country's globally leading mining/industry-based economy, large scale irrigation developments, extensive residential/commercial areas, and international agreements with neighboring countries demand a reliable water supply. Therefore, a system of reservoirs, control structures, diversion facilities, abstraction/return flow structures, and interbasin transfers has been built. Today, it is one of the most complex, integrated, and large-scale water systems in the world.

The Vaal River basin is at the center of this extensive complex that includes linkages to seven adjoining basins, collectively called the Vaal River System. It serves the water needs of South Africa's industrial heartland in and around Johannesburg and also meets downstream agricultural, hydroelectric generation, industrial, residential, and commercial demands in the Orange River watershed.

As this system was progressively developed, operating rules were defined for individual projects and did not fully capture the benefits from operating a fully integrated system. This was especially relevant for the South Africa system with its combination of very challenging demands, extreme hydrologic variability, and enormously complex multipurpose and integrated multibasin system of water regulation, diversion, and control projects.

Our story started when we met during a water resources conference at the University of Waterloo, Canada in 1978. This was the start of a long-standing friendship and has led us to recognize the great need and opportunity in South Africa for the application of "systems methodology" to the planning and management of the country's complex water resources systems. The challenge was how to make it happen.

The custodian of the country's water resources, the Department of Water Affairs (DWA), was a highly professional and capable organization. Virtually all professional work was handled in-house, with minimal use of consultants. It was against this background that we managed to get an appointment with the Chief Engineer (CE) for Planning to share early ideas on the introduction of a "systems methodology" approach to the planning and management of the Vaal River System. After a series of discussions over a period of two years, the CE for Planning was sufficiently confident about the potential benefits of a "systems methodology" approach for the DWA to invite tenders from international consultants to develop or adapt a mathematical model for the overall Vaal River System. The CE for Operations had reservations but was willing to let us try to change his mind.

Our two firms, Acres International of Canada and BKS of South Africa, submitted a joint proposal and were chosen to carry out the work.

From the onset, we realized that the success of the project would be largely dependent on the understanding and acceptance by senior DWA officials, as well as by external stakeholders. Considerable effort was therefore taken to communicate the principles of the approach and the findings/outcomes as they became available.

Initially, presentations were made by the consulting team, as would normally be the case. However, success was enhanced when the appointed DWA project leader increasingly presented results of the project, in person, to DWA's management team.

Early successes from the system analyses included robust assessments of flows and their reliabilities. This formed the basis for assessing how one component of the system could support another in times of need and for development of smart system-wide operating rules, where the performance of the overall system was shown to be substantially greater than the sum of the component parts. Proper understanding of the complex inter-dependencies among components of the system, together with monitoring of actual system behavior, also identified where the operation in practice of some components of the system could be improved.

As the work progressed, it became evident that there were opportunities for further refining system operating rules and for evaluating associated system-wide benefits based on expanding historic hydrological records into a large number of synthetic flow records, using probabilistic methods. This allowed for a more precise evaluation of benefits for a wide range of possible future hydrologic events, especially including very extreme events, with their associated probabilities of occurrence. The next step, therefore, was to demonstrate to DWA the added benefits of using probabilistic methods to the proven system analysis methodology.

Having demonstrated the benefits of a "systems methodology" approach based on historical hydrologic records, we entered into discussions with DWA on the concept of developing a more advanced methodology using stochastic hydrology and probabilistic techniques. Thanks to the professionalism of the officials and the trust in the consulting team, this was approved under a major extension of the assignment. Adding a probabilistic feature to the Vaal River System Model was a major undertaking, given the overall complexity of that system. The probabilistic work was subjected to rigorous testing and international expert reviews, as well as extensive interaction with DWA's management team.

The first critical operational decision with respect to the Vaal River System based on the stochastic models came in 1989, when a decision was taken on probabilistic grounds to lift restrictions after eight years of the most severe drought on record. This decision turned out to be the correct decision. Many other critical operational decisions have been taken since, with resultant savings of millions of dollars to the national economy. One example was a decision to stop pumping for a year at a major inter-basin transfer that conveyed 20 m^3/s of water over more than 500 m head. Savings from this decision, alone, exceeded the cost of our contract to develop and apply models. Even greater savings and benefits were achieved through refined development planning and probabilistic based implementation scheduling.

The initial stimulus for a system approach and methodology was focused on improved operation of the system, by analyzing performance in a systems context (*operational planning*). This goal was well achieved and far exceeded by addition of the probabilistic approach and capabilities. However, as the project progressed, our models were also used increasingly for *development planning* by assessing the performance and feasibility of potential new infrastructure in a system context, while applying real-time probabilistic operating rules. This included selection or rejection of projects, as well as their sizing, scale, and implementation scheduling.

For example, using this overall approach, we demonstrated that the planned last two phases of the giant Lesotho Highlands Water Project would not be economic. Again, major savings were realized from the benefit of applying a rigorous and comprehensive "systems" approach to development planning.

Perhaps the most pronounced vote of confidence came in 1993 when it was announced in parliament that, despite just having experienced the lowest annual runoff on record to the Vaal River System, water restrictions would not be implemented in the central economic heartland of the country for at least two more years. Invaluable user confidence was realized in this way. Again, this decision turned out to be the correct decision.

The systems methodology developed has been accepted as standard practice in South Africa and is still applied regularly for planning and operating the major water resource systems in the country, with ongoing benefits continuing to be realized. It has also had the advantage of creating an understanding of the capabilities and limitations of these water resource systems, thereby creating greater cooperation between economic, political, and environmental sectors. In particular, it has facilitated the building of trust with neighboring countries with respect to the development and management of shared river systems.

The application of systems methodology has also served as an invaluable aid for shifting the overall approach and understanding for development, management, and operations of water resources systems in South Africa, from analyzing performance of individual developments, to developing greater appreciation of overall interaction between projects and for assessing performance and results in a probabilistic broad-based systems context.

Part II

Adventures in Asia

Regulation of Lake Baikal

Mikhail Bolgov, Alexander Buber, and Alexander Lotov

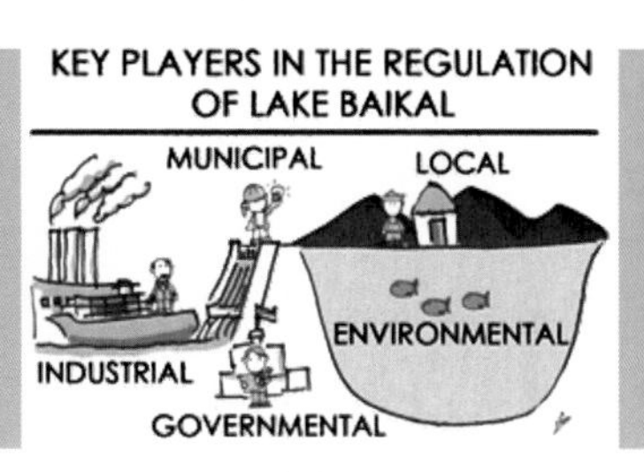

Lake Baikal is the largest freshwater lake by volume in the world, containing more than 22% of the world's fresh surface water. It is located in Russia, in southern Siberia, close to the border of Mongolia. Baikal contains more water than all of the North American Great Lakes together. With a maximum depth of 1,642 m (5,387 ft), Baikal is also the world's deepest lake. It is considered to be among the world's clearest lakes too. Baikal is home to thousands of species of plants and animals, many of which exist nowhere else in the world. Among them is the local fish Baikal omul, which is considered Russia's tastiest fish. Lake Baikal was declared a UNESCO World Heritage Site in 1996. It is also home to local Buryat people, who reside on the eastern side of Lake Baikal.

According to an old Buryat tale, the old man Baikal had many sons but only one daughter, Angara. The sons brought wealth to the father's home, but one day Angara took the wealth and ran away to her beloved Yenisei. Indeed, many rivers bring water to Lake Baikal, but only one river flows from the lake, the Angara River. This river flows into the Yenisei River, which is one of the largest Asian rivers.

In the second part of the twentieth century, a cascade of dams and reservoirs was constructed along the Angara. Now it includes four reservoirs, which are intended for producing renewable hydroelectric energy, for providing transportation through the Angara and Yenisei Rivers, and for avoiding floods, among other purposes. The upper reservoir (Irkutsk Dam) is used to regulate the level of Baikal.

The control rules for the dams have been based mainly on the requirements of the water users. The control rules must satisfy industrial, agricultural, municipal, and environmental requirements. In the case of conflicting requirements, experience of water managers has been used to resolve them. These control rules have not satisfied the people living at the lake or using its water resources. There are conflicts among all of these needs and requirements. Hence, the challenge is to find control rules that provide appropriate tradeoffs among all these interests.

Existing control rules have many disadvantages. One of them is related to the level of the lake. After building the Irkutsk Dam, the water level in Baikal increased about 1 m, which resulted in the erosion of the shore in many places. Moreover, several biologists claimed that variations of the level are harmful for the environmental system of the lake. In particular, they result in decrement of Baikal's omul population. This statement seems strange since the natural seasonal and multiyear variations of the level of the lake (i.e., the variations before the Irkutsk Dam was erected) did not prevent the well-being of the omul population for thousands of years.

Nevertheless, in 2001 the government of Russia decided to restrict the variation of the level of the lake by one meter by setting fixed maximal and the minimal levels of the lake. A scientific substantiation of this decision was not given, much less one that took into account that the natural variation of the level of the lake was about 2 m. The decision became effective in 2003. After 10 years of its implementation, the impacts of this decision are clear. The restriction prevented accumulation of a sufficient quantity of water during the years with high precipitation, causing shortages in dry years. Moreover, the population of the Baikal's omul decreased during these years.

The need for new control rules was clear. We started our studies using multicriteria optimization methods in hopes of finding more effective reservoir Irkutsk release rules. Because we were outsiders for the Ministry of Nature, which was responsible for the situation at Baikal, our study was not supported by the ministry. Our financial support came from a grant from the Russian Foundation for Basic Research (Russian analog of the US National Science Foundation). The grant, however, was aimed at developing new methods for multicriteria optimization. We used the Lake Baikal regulation problem as a real-life application of our methodological studies. However, this application turned out to be more important than we assumed in the very beginning of our research.

At first, we considered the Irkutsk dam alone. We studied conflicting requirements to the control rules, such as the required electricity production, governmental requirements pertaining to the level of Baikal, minimal flows through the dam, prevention of floods in lower pool of the dam, and so on. The number of parameters of the rules (decision variables) exceeded 100. The number of goals or objectives to be achieved was 26, and many were conflicting.

Using our multiobjective decision optimization and visualization methods, we proved that it was impossible to satisfy the governmental lake level requirements over a long period of time. But we also identified, and showed operating authorities, the tradeoffs among flow and level requirement violations when meeting all specified requirements were not possible.

After we concluded our study, a period of low precipitation began. For two years, the inflow to Lake Baikal was extremely low, and the average level of the lake gradually decreased. High precipitation and related inflow was expected the following year, but it did not happen. For this reason, the government of Russia asked experts in water management for help. On the basis of the results of various studies, including our study, a new regulation was prepared and approved by the government. In accordance with this decision, temporarily, the government will decide every year about the minimal level of the lake. It is clear that it will be lower than the level related to the regulation policy of the past.

Recently we received a new grant for continuing our search for improved rules for controlling the level of Lake Baikal. We hope the results of our work will convince the government to adopt permanent control rules that take all water users' interests and ecological impacts into account, while recognizing that tradeoffs among them may be necessary. One of many lessons learned from this experience is that it takes time and patience to change the established ways of doing things. The institutional willingness to consider change is critical for any change to happen. This, in turn, may depend in part on our ability to motivate such change.

Attempting Integrated Water Management in Bhutan

Leon Hermans

We all know water systems impact the lives of many people and link to many other systems that exist in our communities. Thus, to effectively design and manage our water systems, we need to also consider the interactions among all the systems depending on water. The term given to this comprehensive approach is called integrated water resources management (IWRM). One of the current United Nations Sustainable Development Goal (SDG) targets related to water (target 6.5 to be precise) is to implement integrated water resources management at all levels of government by 2030.

In 2006 I was working on IWRM issues for the Food and Agriculture Organization of the United Nations, based at their Headquarters in Rome, Italy. I felt privileged to have a job with this UN organization and to spend some time in Rome. One assignment I had was to work on an IWRM plan for Bhutan. Bhutan is an isolated, landlocked country in the Himalayas.

This beautiful country is known as the "Land of the Thunder Dragon," and flight arrivals at its international airport are among the world's most spectacular. Bhutan is also famous for developing a Gross National Happiness Index as a key development indicator, as a substitute for a narrower focus on the gross national product.

During a Food and Agriculture Organization mission in (i.e., a visit to) Bhutan, I worked with an IWRM specialist of the Royal Government of Bhutan. We had lively discussions, visited various sites in the country, and met with different professionals and stakeholders to give me a clearer idea of the challenges involved in developing a plan for IWRM for Bhutan. As the source of mountain spring waters and glaciers, one might expect that water resources are abundant. Perhaps this was the situation in times past, but not when I was there. Pollution and stream water quality were concerns, as well as climate change and its effects on the melting glaciers and changing patterns in precipitation. Storing water in a mountainous region is difficult, adding to the impacts of climate change. In some watersheds there were tensions and conflicts among water users, for instance between upstream and downstream communities. Deforestation, land erosion, and groundwater management were other concerns. These included the adverse impacts associated with various large hydropower stations, some operational and others under construction, and numerous smaller-scale hydropower facilities for local communities. All in all, there were important challenges and opportunities warranting serious work on IWRM strategy development and implementation.

Any serious undertaking of IWRM planning requires people with skills that allow them to effectively contribute to the planning process. This, I learned, was a limiting constraint in Bhutan. The Royal Government of Bhutan's IWRM expert, who served as my teacher there, was very capable, knowledgeable, and clearly passionate about his work on IWRM. Unfortunately, he seemed to be almost the only IWRM officer in the national government. Moreover, his talents and skills did not go unnoticed within the government system. Soon after our mission in Bhutan, he was assigned a new position at the more prestigious department of transportation and road construction. This would leave the position of IWRM coordinator vacant, with no good specialist readily available.

Clearly, the absence of sufficient resources, especially skilled experts in various positions within various organizations, makes the implementation of IWRM very difficult. Yet the events in Bhutan were not so strange, and are not unique. Bhutan has a population of almost 800,000 people. The literacy rate is slightly above 55%. And water has, historically, not been the biggest challenge for the nation's development. Obviously they need, and have, engineers able to design and operate water supply and sanitation facilities, storm water drainage designs, reservoirs, and hydropower plants. But while we might expect every country to work on IWRM and have IWRM specialists, we do not necessarily expect every medium-sized municipality to do the same. In terms of population, Bhutan, like some other countries, compares to a medium-sized city.

As water experts, we tend to emphasize and give priority to water systems, but there are many issues affecting other systems that compete for attention. Many outside the water sector often take water for granted and focus on other societal challenges. We should not forget that for most people, governments, and businesses, water problems may just be several among many. And if we do forget, we will surely be reminded at some point—as I was in Bhutan.

Political Challenges of a Water Expert in Iran

Pete Loucks and Laurel Saito, editors

"I guess my environmental work is not welcome here."

This is a story about Kaveh Madani, a young and highly respected professor at Imperial College London, who returned to his homeland of Iran with "the hope of creating hope." It's a brief summary of adventures of a patriotic scientist in a developing country facing massive environmental disasters. It is a story of how good science that can bring about big positive change in real life was not tolerated by powerful hardliners who preferred shortsighted, business-as-usual routines. Kaveh should be writing this, but you will understand why he isn't as you read on.

Iran is currently struggling with a host of environmental issues including acute water shortage problems, drying up of lakes and wetlands, desertification, dust storms, and rising competition and tension about how to share limited water. Kaveh Madani's work on addressing environmental problems in his home country, Iran, has been featured in the major news media as well as in our profession's more technical literature. His research and consulting accomplishments and reputation as an advocate for improving the management of our natural resources and environment are widely acknowledged, including by the ruling political administration in Iran. In September 2017 the Iranian government persuaded Kaveh to leave his prestigious position at Imperial College London where he taught systems analysis and policy to serve as the Deputy Head of Iran's Department of Environment (DOE). This made him one of Iran's Deputy Vice Presidents and the second most powerful politician of the country in charge of environment. During only seven months on the job between September 2017 and April 2018, Kaveh made remarkable contributions by increasing public awareness and creating synergies to solve environmental problems. His work earned him the praise and respect of his fellow countrymen inside and outside Iran.

Although an Iranian citizen, Kaveh was considered an outsider in Iran's closed political system. Like many bright Iranians of his generation, Kaveh left Iran after getting his bachelor's degree in civil engineering. He got his master's degree in water resources at Sweden's Lund University and his civil and environmental engineering Ph.D. at the University of California, Davis. Kaveh then conducted postdoctoral research in water resources economics and policy at the University of California, Riverside. He left Riverside to become an assistant professor at University of Central Florida before being hired by the Centre for Environmental Policy of Imperial College.

Kaveh quickly gained a global fame for his work on decision-making, game theory, and water management. He won prestigious awards from the European Geosciences Union (EGU) and American Society of Civil Engineers (ASCE). His reputation grew further after his popular TEDx talk about global water conflicts and appearance in documentaries on water and environmental problems in Iran and the Middle East.

While he worked outside Iran, Kaveh kept in touch with his homeland. He traveled home regularly, sometimes to lecture, other times to run conferences or consult. He attracted the attention of Iranian journalists, consistently on the lookout for successful fellow Iranians abroad. He emerged as a frequent commentator on the environmental crises Iran faced. Speaking in a straightforward, accessible way, he established a firm online national and international presence and used catchy phrases like "water bankruptcy" instead of the more commonly used "water crisis."

President Rouhani's administration had promised to give top jobs to innovative, forward-looking professionals, and not necessarily restrict itself to hiring people who had done well within Iran's political establishment. The government took note of Kaveh's record as an internationally acclaimed water scholar, and his approach to his work.

In September 2017, Kaveh was brought in by Mr. Issa Kalantari, vice president of Iran and the current head of Iran's DOE and a former Minister of Agriculture, not only to address Iran's water crisis, but to set an example for other Iranian expatriates to return to Iran. He was seen as a symbol of expatriates' return to Iran and President Rouhani's efforts to reverse Iran's brain drain issue. Tehran announced Kaveh's appointment, making him the deputy vice president in charge of environment. This was a historic milestone for the Islamic Republic of Iran after the 1979 Islamic Revolution and marked the rare return of a popular and eminent Iranian expert from the diaspora who did not belong to the inner circles of the Islamic Republic. Kaveh's appointment came with a huge positive public reaction and excitement. He had become a symbol of expat Iranians' return, a role he was very aware and perhaps excited about. "There are a lot of people abroad, waiting and watching closely to see what is going to happen," he tweeted after being appointed. "If I succeed, we might see more people coming back to help the government."

In an interview after his appointment, Kaveh said that his parents and friends had been against him moving to Iran, as were many Iranian expatriates on social media. "They all said that the government would bring a man of science into an executive position and then destroy him." Such words seemed too pessimistic at the time, but they proved prescient when Kaveh was arrested in a crossfire between the Iranian hardliners and the more moderate ruling administration.

Upon Kaveh's return to Iran, intelligence agents from the Revolutionary Guards stopped him at the Imam Khomeini International Airport to question him at length about his intentions of going back to Iran. They hacked into his email and social media accounts and electronic devices to access his personal and professional information. He was constantly under pressure by the Iranian radical intelligence forces, who never trusted him and he was interrogated many times by them during his tenure in Tehran. A few months after his return, he was detained for three days when environmentalists were arrested in a wave of being accused of "environmental espionage." Those arrested were charged with acting under orders from the CIA, MI6, and Mossad to infiltrate the Iranian government and gather information about the country's water situation and sensitive military sites, including missile installations. Although Kaveh was released in response to strong international and national reactions that made Rouhani's administration accountable for his situation, other environmentalists have remained in jail and one of them, Kavous Seyed Emami, "committed suicide" in prison. Seyed Emami's alleged suicide was doubted and debated by various human rights organization who thought it hard to believe.

Kaveh Madani became a victim of Iranian authorities' hostility toward educated Iranian expatriates, many of whom hope to return to Iran to do good. The Iranian Revolutionary Guards' intelligence forces accused him of "creating a dark image of Iran's water situation." They called him a "water terrorist" who had been trying to justify the shutdown of the agricultural sector to force farmers to migrate to cities and create tensions and political conflicts. He was also labeled as a "bio terrorist" who had plans to make the country food-dependent on foreign forces, import GMOs, and biologically destroy the next generations of Iranians. Other accusations aimed at him included "trying to ratify the Paris agreement on climate change" and "cooperation on realization of UN Sustainable Development Goals (SDGs)," claiming such efforts could limit Iran's development!

Kaveh had to finally resign because of unrelenting pressure from the Iranian hard-liners and the hostile campaigns launched by security agencies to damage his image publicly and harass his family. Their efforts were apparently enough to finally convince Kaveh that he would not be tolerated by those who are threatened by good science and patriotic efforts to save Iran's natural resources. In April 2018, Kaveh resigned and left his homeland forever. "Yes," he wrote on Twitter, "the accused man fled a land where ignorant trolls rejected science, knowledge and expertise and with the delusion of conspiracy were in search of someone to take responsibility for all problems because they know well that finding a blameworthy, enemy, or spy is easier than taking responsibility and joining forces to solve the problems."

His resignation was followed by dramatic emotional reactions by the Iranian public and the expatriates who interpreted the course of action against him as clear evidence of how a gifted Iranian patriot returned with good will and tried hard to address the environmental problems that had been created by his peers back home. While it is a tradition in Iran to blame problems on those in power, Kaveh Madani didn't do that. He told the Iranians that they all bear responsibility for the problems and they are all bound to push for progress. Iranians who really cared about the Land of Flowers and Nightingales had to be willing to fight for what is right. Individual hypocrisy was the enemy, in his mind.

Kaveh is still active on social media, trying to raise environmental awareness and to capitalize on his popularity and fame to get the Iranian public engaged in addressing the country's serious water and environmental problems. Kaveh was Iran's second man in charge of the environment for a relatively short time, but his science, popularity, and approach helped him accomplish a lot according to his colleagues, environmental activists, and the media. In response to his resignation letter, Vice President Kalantari praised his work by writing "without despair, you welcomed the opinions of the public and experts with a positive attitude. You will never be forgotten by the Department of Environment's personnel or the people of Iran. Given your brilliant capabilities and competence, as well as your intellectual maturity and firm belief in the national interests of this country, I am certain you will never deny helping the people of Iran wherever you may be."

Khong, the Mother of Water

Leif Lillehammer

I joined my current workplace, Multiconsult, in autumn of 2013. In the spring of 2014, I worked on a proposal for the Mekong River Commission (MRC). We were awarded the project later that year. The Mekong had always been a mythical river for me because of my fascination with the old Khmer civilization as well as the movie *Apocalypse Now.* I heard the propeller blades of the helicopters of The Doors song "The End" (the theme song of *Apocalypse Now)* in my head when I realized we won the contract, but with an opposite meaning—this was not the end but rather a new start. The project lasted for over three years, ending in May 2018. Through a number of studies, including some detailed hydrological-hydraulic-geomorphic-hydropower modeling all focused on identifying best-practice mitigation approaches to hydropower development, we have now helped MRC launch their Hydropower Mitigation Guidelines. This is a technical guide that we expect, and hope, will support other initiatives of MRC and their member countries.

The project took me to Laos, Cambodia, Thailand, and Vietnam 13 times over these three years, including a field trip along, and on, the Mekong River, also known as the Khong, the "Mother of Water." The latter was an epic journey that also included multiple dawns and sunsets along this mighty river. This will be etched forever in my memory.

This project in the Mekong Basin sparked an international event in my hometown of Oslo in September 2017. The Oslo Hydropower Sustainability Forum was an event jointly organized by MRC, the Deutsche Gesellschaft für Internationale Zusammenarbeit (GIZ) GmbH, and Multiconsult here in Oslo, with additional support from the Norwegian Agency for Development Cooperation (Norad) and a Dutch engineering firm Deltares. Participants came from all around the world, dialoguing and discussing "New

Frontiers in Hydropower Development." My Cornell University engineering colleague, Pete Loucks, participated by presenting, among others, a keynote speech on the results of a parallel project on hydropower development in the Mekong Basin and its potential impact on fish migration and sediment transport. Imagine, an engineer talking about fish migration and survival! Almost 20 years had passed since we first met at Cornell, and somehow, our interaction continues.

Enjoying Caviar in the USSR

Pete Loucks

"Whether capitalist or communist, managing water is still a challenge."

From the late 1960s to the late 1970s, there was a thawing of the ongoing Cold War between the United States and the Soviet Union. This détente took several forms, including visits of Soviet scientists to the United States to observe how we do what we do, and visits of US scientists to the USSR. One such series of exchanges involved the US Environmental Protection Agency and the All-Union Scientific Research Institute for Water Protection (VINIVO). I was one of three participating in those exchanges focused on how each country managed the quality of water in its rivers. Our translator was someone that had learned English only by listening to BBC, but his English, even slang expressions, was excellent. The only expression that we discovered he didn't know was "go bananas." I recall VINIVO telling us, at the beginning of our visits, that they believed the reason why we capitalists could not implement cost-effective policies for controlling wastewater discharges and river water quality was that the municipalities, industries, and farms creating the polluting wastewaters were not under the control of one governmental agency that could impose such policies. However, those in the USSR, they said, can implement such policies because they are under a centrally planned and controlled economy.

Well, after a number of exchanges, over a number of years, involving lots of caviar, vodka, music, hospitality, and interesting new experiences (some involving those responsible for security), our Soviet colleagues had to admit that they had the same problem as those of us living in our capitalist societies. Even under a centrally planned economy, each minister of each agency overseeing the activities of its sector of the economy had their own objectives and budgets. Interagency coordination permitting cost-effective approaches to pollution control was not a prioritized part of their mission. Human behavior seems to be the same, regardless of what economic or political system we live under.

As a postscript, while I was working in Poland, I learned that in an attempt to reduce the discharge of industrial wastewaters into rivers, the Polish government levied a fine on firms whose discharges were excessive. Since the government owned each firm, they were paying the fine to themselves. All each industrial firm did was to transfer their fines from a column on one page of their account ledger to another column on another page. The fines had no impact on their production and resulting wastewater discharges. Eventually realizing this, I'm told, the government then required the employees of the polluting firms to pay the fine. That motivated these employees to put considerable pressure on their companies to reduce their wastewater discharges. I was told that policy worked. Clever Poles.

Growing Qat in Yemen

Pete Loucks

"If I have to give up my qat, you should give up your coffee."

While working in Yemen on a project devoted to finding ways of prolonging the life of the aquifer that the capital city of Sana'a depends upon for meeting its water supply needs, we discovered the biggest demand on that aquifer was for irrigating a crop called qat (or khat). Qat is a narcotic, and its use is supposedly illegal. However, just about everyone in Yemen enjoys chewing it. Growing it takes a lot of water. For us, an obvious way to extend the life of the aquifer was to restrict the irrigation and consumption of qat. This suggestion would be similar to asking everyone who drinks coffee to stop drinking it. Other than fixing leaks and preventing theft, we were not able to think of any major way to reduce the withdrawal of water from the aquifer. However, we did suggest that the government begin a public discussion of the problem. This could generate, in time, greater public awareness and support for more restrictive water-use policies. The government's response to this suggestion was to do the opposite, citing that by the time the aquifer became unusable, they, the politicians, would no longer be in power, if indeed even alive. Perhaps they were right. But certainly one of the causes of today's conflicts in Yemen, some decades later, stems from their shortage of water.

Peace Talks on Water in the Middle East

Jerome Delli Priscoli

In the early 1990s, I helped write the briefing books to prepare the US Secretary of State for the Middle East negotiations that took place in an effort to achieve peace in that region of the world. One of the five major issues, or tracks, involved water. The United States became the gavel holder for this track, and I served on the US negotiating team for four years. Water had become a venue for building cooperation, trust, and confidence. Actually, water had already been such a venue in the Middle East. Even as the parties were forbidden to talk in public, they met continually in the so-called table talks along the Jordan river.

I remember the tension in the room in Vienna, Austria, as I delivered the opening talk on water. As I looked around the room, however, it was like old home week for me. Around the table were familiar faces and close colleagues. That was the first time I actually experienced the reality of water as a tool in public diplomacy. We were talking about serious issues with huge consequences with representatives of peoples who had been killing each other.

The second time such reality hit was at a reception in Washington, DC. A Palestinian representative was speaking with two Israelis. They asked, "How much water do you really need?" A number was floated, and the Israelis said this should be easy to divert from the Israeli national carrier. The Palestinians retorted that they did not want that water; they wanted their water. In short, they wanted to see certain physical water from specific places that symbolically belonged to them—water that they viewed as stolen from them. What a lesson about the powerful symbolism of water, about the need to find some rituals of mutual acknowledgments of hurt.

During one acrimonious period, the parties (under orders from their governments) sat around the table for a day saying virtually little to nothing. However, at the breaks they would gather and draw out solutions to regional water issues on pads of paper! The following day we arranged a private plane to fly to the Tennessee Valley Authority operations center in Knoxville. In the morning at the airport, no one was talking. Twelve hours later, on the way home, the plane sounded more like one of the local sports bars. This group of high-level technopolitical experts had just spent the day engaging in what they knew and loved—together! To this day I still hear some of the participants' talk of that plane ride back to DC.

Obviously, things were not always rosy. We came close to historic agreements several times only to fail at the end. But the talks did establish some things. In fact, 10 years later I was asked to attend a lunch at the Dutch ambassador's residence in DC. The topic was how to start talks after the intifada. Why was I there? It turns out that the only existing organization in the region that had all the parties as signatories was the regional desalination center created by the water multilaterals! The Dutch and others thought such an organization was just what was needed to bring parties into dialog.

It is impossible to ignore the importance of public water diplomacy after such experiences. The multilaterals used water as a venue of dialog to bring together parties who otherwise were not talking. This is but one case. There are many others, such as the Indus Waters Treaty and how it has held through periods of violence.

Combating Water Scarcity in Jordan

David E. Rosenberg

"This shuts off the water after 15 minutes. Perfect if you have teenagers."

The Hashemite Kingdom of Jordan is one of the most water-scarce countries in the world. Water use far exceeds renewable supplies with the deficit met by unsustainable pumping of groundwater. Water-level drawdowns in wells commonly exceed 1 m per year. Although nearly every household in the country is connected to a potable, piped water-distribution system, these systems are typically pressurized for less than 12 hours per week—that is, less than 7% of the time. Jordanian households therefore typically store water in rooftop and cistern storage tanks. Water use in Jordan averages only 100 L (26 gal.) per person per day. Many people have to make do with less. In a country where water is so scarce, what role can, or should, water conservation play in integrated water resources management?

My introduction to Jordan's water issues began while I served as a US Peace Corps volunteer there in the late 1990s. I worked at the Azraq desert oasis wetlands reserve located on the main highway between Amman and Baghdad. Azraq's water challenges were a microcosm for Jordan's national water challenges. Nearby farmers in the surrounding desert pumped groundwater to grow low-value tomatoes while the government pumped from the same groundwater aquifers to supply major urban areas to the west in Zarqa and Amman. After the springs that fed the wetlands went dry, the reserve installed its own well to pump water from the aquifer and rewet the wetlands. Groundwater levels continued to drop. I worked on reservoir habitat monitoring, education, computer training, and composting programs, and I learned Arabic. Speaking Arabic helped me understand and navigate the interpersonal dynamics among the reserve staff as well as talk to neighbors, friends, and strangers about their water problems.

Jordanians love to talk about water! While I bought eggs and hand soap, the underemployed guy who ran the corner store near my house held a master's degree in groundwater hydrogeology from the prestigious Jordan University of Science and Technology. He described to me the aquifer recharge zone 50 km north in Syria and the flow paths that created the Azraq springs and wetlands. On the bus or in taxis, people told me how Jordan's neighboring countries to the north, west, and southeast did not uphold their water-sharing agreements. People claimed that the water that did come through the distribution system was contaminated. The recent public-private partnerships created to more effectively manage urban water charged more money for worse service.

Between 2003 and 2005, I returned to Jordan eager to apply my recent graduate education in integrated water resources management. I wanted to identify new supply and water conservation actions that individual households, the Amman water utility, and Jordan as a whole could adopt to balance supplies and demands. I proposed to build and run a model to identify the cost-minimizing mix of long- and short-term actions to achieve this balance across a set of summer and winter uncertain-water-availability events.

New supply actions for households included installing roof, ground, or cistern storage; installing rainwater or graywater collection systems; or buying water from private vendors. On the conservation side, households could install water-efficient showerheads and faucets, low- or dual-flush toilets, automatic laundry machines, and drip irrigation systems; place bottles or bags in toilet tanks to reduce the flush volume; fix leaks; reduce shower time; or sweep tile floors rather than wash them, among others. I drew on lots of prior studies to estimate costs for each action and how much water each action would add or save. Where information was lacking, I visited stores or the individuals who sold the products, posed as a customer, and negotiated a price to purchase. Nearly all Jordanian stores display a marked sale price as the *starting* (asking) price. The *actual* price depends on the buyer's bargaining skill and patience.

I had multiple values for each cost and water-use parameter and used these values to construct probability distributions that described the likelihood the cost and water-use parameter would take a particular value. Using these distributions, I developed a set of 500 simulated Ammani households. At about the same time, I acquired from Amman's water provider the distribution of piped water use for the entire Ammani population. I used these data to calibrate my household model. Household occupancy was uncertain because between 2003 and 2005, hundreds of thousands of Iraqis had fled Iraq and taken up temporary residence in Amman because of the US–Iraq war. Jordan was no stranger to refugee waves. Similar waves appeared in 1948, 1967, the 1980s, and 2013, connected to two Middle East wars. For example, Jordan currently hosts more than 600,000 Syrians.

With my calibrated household model, I was off and running to identify how households could save money and water. Most Ammani households used a very small volume of water per quarter. However, a sizeable subpopulation of households used much more. This meant households had different potentials and abilities to conserve water. Subsequently, I have seen similar water-use patterns in many other cities, suggesting that water managers should target conservation actions to the high-consuming households that have the most potential to save water and money.

At this point, my work had focused on simulated households. What insights and surprises awaited if I looked at actual households?

My Jordanian friends recommended families who would be willing to meet with me to discuss their household water use. During such visits, I asked about the availability of their piped water service, storage capacities of roof, ground, and cistern tanks, drinking water practices, and whether they would be willing to implement new actions. Then, I finished each visit by asking the family to suggest other families I could visit. Social scientists describe this sampling strategy as snowball sampling. I found it slightly ironic that snowball sampling worked in the hot, dry, desert of Jordan.

In visits with households, I found it very important to ask questions in different ways and look around the house. Not everyone saw or grouped the water supply, conservation, and reuse actions as I did. For example, one person responded "no" when I asked if they reused water. However, I later saw inside their guest bathroom a bucket that was used to catch leaks from a leaky shower head. When I asked what they did with the water in the bucket, I learned they used it to mop floors and fill their semiautomatic washing machine.

It took a few weeks to organize the responses I heard during the visits. As I was doing this, I had to return to the United States. Several months later, after running my model for each house, I returned to Jordan to share the results.

Most of the people I originally visited were excited that I had returned. They were used to foreigners collecting lots of information and then disappearing. In each second visit, I listed the new supply and conservation actions the model recommended the household should adopt, and I solicited reactions. Several families said they were willing to try actions the model recommended, such as putting bottles in toilets to reduce the flush volume or installing a dual-flush toilet. Other families indicated that in the intervening months they had already adopted some recommended actions like recycling graywater, collecting rainwater, using a car wash, or installing a countertop water-treatment system. Still other families described reasons they would not implement an action the model recommended (e.g., they had a maid, drip irrigation is for bigger farmers, rain water has poor quality). At the end of these visits, I asked a final open-ended question—what other conservation actions were they thinking about besides ones we had discussed?

One father and head of household had a surprising answer. He and his family lived in a multibedroom apartment in a well-to-do, high-story building. For both the original and follow-up visits, we sat in a plush living room with a fancy entertainment system. We snacked on apples, bananas, cookies, sweets, and tea that his wife regularly brought in. Otherwise, his wife stayed out of sight. To the final question of what other conservation actions he was considering, he answered:

"I am waiting for my oldest teenaged daughter—who takes really long showers—to get married and move out of the apartment." From the next room, his wife laughed in agreement. The comment was both funny and a bit intimidating because it highlighted several deficiencies in the water conservation research and work.

First, there are substantial differences in water use even among household members, let alone across households. Second, heads of household cannot always control the water-use behaviors of their family members. And third, water conservation—like irrigation efficiency—is quite context specific. His daughter's marrying and moving out may reduce his household's water use. But the city will see that same water use in the new location where she and her husband decide to live. Such "water conservation" may just move a water problem from one place to another. And today, despite higher-frequency temporal data on water use, more sophisticated models, and surveys designed and carried out by credentialed social scientist collaborators, my students and I still grapple with the full implications of water use, water conservation, and other behaviors impacting water system management.

Building Collaborative Networks

Laurel Saito

"I couldn't have gotten this far without my team."

In 2003, as a new professor at the University of Nevada Reno, I took advantage of an opportunity to travel to Uzbekistan in Central Asia. I wanted to visit some friends who were working on a water resources management project. During this visit, I met a medical scientist with the Uzbekistan Academy of Sciences, Dr. Dilorom Fayzieva. She specialized in water quality and health issues. Someone suggested that we apply for a grant that would enable us to work together on these issues. The American Association for the Advancement of Science had a program that offered funding to help female scientists from different countries to meet and prepare proposals for research funding. We applied and were given funds that allowed me and a colleague from the US Geological Survey (USGS) to travel to Uzbekistan to meet with Dr. Fayzieva and several of her colleagues. One of her colleagues was from the University of Bonn and was working on a German-funded UNESCO project on land and water management in the Khorezm province of Uzbekistan.

Dr. Fayzieva, my colleague from the USGS, and I decided to apply for a grant from the North Atlantic Treaty Organization (NATO) Science for Peace Program. We proposed to collaborate with the Germans on assessing the economic value of hundreds of small lakes in Khorezm. Khorezm is part of the Aral Sea basin where irrigation diversions for cotton agriculture have led to the decline of the Aral Sea and the increase in many adverse health and environmental impacts. The hundreds of lakes in Khorezm occur in shallow depressions in the landscape and collect irrigation canal and drainage water. It was thought that aquaculture or other economic uses of the water in the lakes might be possible to provide another alternative to growing cotton in the region. Dr. Fayzieva and I submitted a proposal to NATO and were invited to interview at the NATO headquarters in Brussels. We were told we were the first female partnership to apply for a NATO grant. We were selected for four years of funding for the project in 2006—three years since my first visit to Uzbekistan.

The partnership with Dr. Fayzieva was essential to getting the funding as well as successfully completing the project. She was very good at recommending other scientists who could contribute. One of my students was able to get a Fulbright scholarship that allowed her to work with Dr. Fayzieva along with four Uzbek students. Two of the students completed Ph.D.s based on the project, and others getting their bachelor's and master's degrees also benefited from the work.

Dr. Fayzeiva had a lot of experience with international projects, so she knew some of the challenges of getting international funds and equipment to Uzbekistan. Because of our affiliation with the German project and because we were foreigners, we were able to open a personal bank account with our Uzbek colleagues as signatories. In this way, we were able to use project funds to pay project expenses. Any funds placed in government agency accounts were not likely to be seen again.

In 2005, diplomatic relations between the United States and Uzbekistan deteriorated. Consequently, it became very difficult for us to get visas to travel to Uzbekistan, although we were able to use our relationship with the German UNESCO project to apply for, and obtain, them. This was critical for the successful completion of our project. I was told that during the period of 2006 to 2008, we were the only US-based science project in Uzbekistan.

Another advantage we had because of our collaboration with the German-Uzbek project was the use of their facilities. The German project had an office building with laboratories for storing and analyzing samples. They also had drivers who could take us to the lakes, many of which were very remote and hard to find. The Germans also had guest houses in Urgench and Tashkent, which enabled our students to spend several months in Uzbekistan at low cost and with easy connection to the laboratory and transportation facilities.

Our Uzbek partners also were essential in enabling us to collect data in collaboration with villagers in Khorezm. Because we did not speak Uzbek or Russian, we needed our Uzbek partners to translate. We also relied on them to advise us regarding the social and political implications of our interactions with Uzbek villagers. In some cases, the Uzbeks were cautious about interacting with foreigners, and we needed to respect their concerns.

Through this project, I learned how important it was to have strong partners, especially for international projects such as this one. With strong partners, we were able to develop a solid team of students and scientists to carry out the project successfully. In almost all endeavors we undertake, our success will largely depend on the network of colleagues we are able to build and maintain.

Working with Farmers of the Sihu Basin(四湖流域), Hubei, China

Slobodan P. Simonovic

"Your farm flooded? Turn on your pump and we'll handle the rest."

I recall sitting over dinner in my hotel room in Wuhan and thinking about the long day behind me. An amazing experience. I spent the whole day working with Chinese farmers and got them excited about multiobjective decision-making. Imagine that!

This work was part of a project to provide flood protection and prevent the waterlogging of agricultural lands in Sihu. The project objective for my team was to improve the existing Sihu water-management technology through the development of a hydrological model, a hydraulics model, an operation planning model, a system simulation model, and a real-time operation model.

The Sihu basin is situated in the south-central part of Hubei Province. The basin's population is about 4 million. The area suffers from surface flooding caused by the surrounding rivers, by rainstorms, and from surface waterlogging of agricultural lands due to poor drainage of the runoff. The Sihu water resources management system includes an extensive dike system for protection against flooding caused by the surrounding Yangtze, Hanjiang, and Dongjinghe rivers. It also includes a large drainage system consisting of six main canals, two large storage lakes (Changhu and Honghu), seven main sluices, 17 first-stage pumping stations, and several hundred second-stage pumping stations. They were all in need of repair, rehabilitation, upgrading, and modernization.

The Hubei Water Resources Bureau was planning to use a World Bank loan for improving the engineering infrastructure and operations facilities. In addition, the bureau initiated improvements in system management as an initial step in the development of a comprehensive plan for sustainable development and management of the water resources in the Sihu basin. The Jingzhou Prefecture Flood Protection Office manages the Sihu drainage system. The Sihu Engineering Management Commission in the Jingzhou Prefecture is the basin water resources management unit in charge of engineering works and their maintenance. The required information for decision-making includes recorded rainfall, water levels, and discharges; climate and runoff forecasts; operating conditions of drainage facilities; instructions issued by the provincial flood protection office; and feedback suggestions provided by the operators. The information is sent to the Prefecture Flood Protection Office and its Hydrological Information Group by phone or electronically. The group processes the information for distribution to decision-makers and the operations dispatch group. The dispatch group proposes options for operations based on available climatic and hydrological data, established operating rules, and the intentions of the decision-makers. The final operations decision is usually made in a regular meeting and is based on the proposed options and other considerations. The decision-makers or the prefecture administrative officials chair the meeting. The final operations decisions are sent back to the county and city flood protection offices, as well as the Sihu Engineering Management Commission for implementation.

The ultimate objective of the Sihu management study was to develop an advanced decision support system for guiding the real-time operations of the Sihu basin drainage system. The main part of the proposed approach included the development of a planning model and a real-time operation model. The planning model is designed to generate system operating rules. These rules are used in a real-time operations model. The real-time operations model is designed to identify optimal operating decisions based on the current state of the drainage system and a hydrological forecast that is used to characterize and capture the short-term variability of the hydrological input. These operating decisions are then considered by the drainage system manager for implementation in real time. Implementation involves the cooperation of local farmers who operate hundreds of second-stage pumping stations.

My day started in a training session with about 15 farmers (representatives of different regions) together with the representatives of the Hubei Water Resources Bureau, the Jingzhou Prefecture Flood Protection Office, and the Sihu Engineering Management Commission. The first part of the day was devoted to introducing the multiobjective decision-making process and a modeling approach that could support this process. The main challenge was to present the concepts clearly without being too technical. The translator was working hard to choose adequate terminology in Chinese language. By lunchtime we had successfully finished the first part, and I had to respond to many questions. The afternoon was planned as a real exercise where all the participants had to help select an operating strategy based on their own preferences and the current and forecasted conditions in the drainage system. Participants had to select a set of objectives, assign their preferences, and compare available alternatives accordingly. I was expecting that the participation of different groups (flood protection office, engineering management commission, and farmers) with clearly opposing priorities would turn the exercise into a serious confrontation.

After my presentation of goals for the exercise and procedure to follow, the discussion started—in Chinese. The translator was not able to catch everything, and I slowly lost the ability to follow. The room temperature was raising, the voices became louder, and a lot of talking was followed with fists hitting the table. After about three hours, the work was done, and with the help of the translator, I was presented with the outcomes of their discussion. To my surprise, the participating parties reached complete agreement in selecting objectives and assigning their preferences. Consensus was reached even before the multiobjective analysis was performed. The smiles on their faces reflected real satisfaction with a job well done and a solution that everyone was able to accept. The day finished with some friendly after-work talk and Chinese beer—most of the Chinese were drinking Budweiser beer, and I was drinking Tsingtao.

What did I learn from participating in this project? The efficient and optimal management of complex drainage systems is important. Minimizing flood damage is the most important objective in the Sihu basin of China. Decision support tools, including system optimization models, are useful for identifying operational decisions. Complex decision-making processes require technical support. The involvement of decision-makers at all levels improves management. Training and institutional development are essential components in the practical application of models for identifying and evaluating management strategies. In order to build an efficient decision-making environment, knowledge transfer and the collaboration of local stakeholders are essential.

Note: In 2003 the Sihu basin project received the Canadian Consulting Engineering Award of Excellence, Category: International.

Restoring the Marshes of Mesopotamia (Part 1)

Eugene Z. Stakhiv

"We bombed this place, now we need to rebuild it."

When President Bush declared the end of the war with Iraq, I was asked to help with the reconstruction. I volunteered, along with thousands like me, to help Iraq's people and its 24 ministries recover from the shameful neglect of Saddam's regime and from the consequences of the war. The primary task was to reestablish a functioning Ministry of Irrigation, and the secondary was to restore their "Garden of Eden"—the waterworks of an ancient hydraulic civilization, the marshes of Mesopotamia. I responded in part from patriotism and in part from romanticism. As a child of Ukrainian political émigrés to the United States, I understood how the Iraqis felt about the liberation of their country. I was honored to be part of the first group of individuals to play a lead role in assisting Iraq. I was appointed the senior advisor and de facto interim minister, working first for Lieutenant General Jay Garner and the original Office of Reconstruction and Humanitarian Assistance (ORHA). Most of our Department of Defense and State Department at that time expected we would be out of Iraq in six months. We were asked to prepare a six-month plan for getting the ministries functioning again before we set out for Iraq.

My first meeting with Jay Garner in late April 2003 was strange because he picked me out of a group of advisors and declared, "Are you the guy who's going to take me to see the marshes?" In the midst of our reconstruction chaos, it was a sign of how important the marshes were going to be in the overall reconstruction strategy. I knew before I left the United States that the marshes had an international constituency and that restoring them would involve addressing many complex technical, social, and cultural issues. We were going to have to prepare the ministry to address these issues along with a myriad of other high-priority tasks of democratic "nation-building" without functioning hydraulic infrastructure or even a phone, email, or an office to work in.

As to the romantic part of my decision, I viewed my brief time in Iraq as a once-in-a-lifetime opportunity to breathe the dust of ancient civilizations; examine the ruins of Babylon, Ur, and Nineveh; and travel along the ancient trade routes and paths of classical history. Mesopotamia was at the crossroads of millennia of relentless and ruthless invasions and conflicts. There was no better way to understand the basic nature of these people who, more than most, had their civilization shaped by a harsh environment, countless invading armies, and hundreds of years of occupation by one satrap or another, reflecting the need to control the people and the environment for basic survival.

We came to Iraq with a confidence in the Iraqis' abilities and their capacity to convert their country to a modern, first-rate, first-world nation. We were there to show them how to set up businesses, introduce a market economy, reform the ministries, and make Iraq the model for the rest of the Arab world. In my mind, Iraq could have easily become the California of the Middle East—it had the technological base and an agricultural base as the platform for revitalization after Saddam. But they couldn't move forward without a stable government, and this, we were to learn, would prove very difficult to reestablish. Iraq experienced the destructive equivalent of a tsunami after the invasion of the coalition forces, but it was not the military who caused it; the populace went on a looting rampage that set back our reconstruction timetable. We from the United States were naïve about the chaotic political situation that we encountered.

It was both a war and an economic collapse, topped by the pent-up frustrations of millions who had suffered under the oppression of Saddam for 30 years, transformed into rage and a rampage of looting and pillaging that followed the invasion. It was ten times as destructive as the bombing. On top of that, in our typical brash American way, we were trying to inject democracy into a tribal culture that resisted even small changes in their governance.

Saddam's army had mined several of the large dams, but to their credit, the Iraqi military refused to detonate them. Before the war, I was on a team of specialists who developed models and information for the military about the degree of flooding to be expected if Saddam destroyed the reservoirs and barrages to impede the invasion force. I was happy to see that this infrastructure, which is vital to the lives of all Iraqis, remained intact. Like our soldiers, who feared the possibility of biochemical warfare, I was afraid of being expected to create miracles and provide immediate water services without that infrastructure.

Fortunately, I had two assistants: a US Army Corps combat military engineer, who had a degree in hydrology from Cornell, and a Persian American civilian hydraulic engineer. My combat engineer commandeered an entire Arkansas Army Reserve unit that made it possible for us to travel and do what we had to do around Iraq. We had to visit each dam, barrage, and major pumping station to assess the damages and needs. In addition, that Army Reserve squad from Arkansas became the enthusiastic trainers of a 1,500-man Iraqi security force—the first such police force among all the ministries. Without them, we could not have achieved half of what we did in the first few months, for they protected important equipment supply depots around Iraq that were used to maintain existing equipment, pipelines, and pumps.

In the first month, we began the slow, systematic process of assessing damage, setting priorities, paying the workers, and repairing hundreds of pumping stations, dams, barrages, and offices that were damaged and looted. We did a lot of scrounging and begging for simple pieces of equipment, and with the military—one of the most infuriatingly cumbersome bureaucracies around—everything was a struggle, but it was the only show in town at that point. They were in control of the hinterlands, and we had to conform to their rules. The pace was relentless. I was working 18 hours a day, seven days a week with a skeleton crew, barely making a go of it. I finally understood what "catch 22" meant—many times over. It also happened to be the first of over 100 straight days of 100+ degrees (F) in Baghdad, and hotter yet in the hinterlands. By August, the month that the Iraqis called "the furnace," we prayed for a "cool" 100-degree day. I remember one day the forecast was for a cold front to pass through Baghdad; the temperature dropped to 104°F at midnight, when I was returning from the office to my quarters. We felt the relief—it actually felt pleasant!

There were massive problems of looting, and Iraq needed every kilowatt of electricity generated at the 10 dams operated by the Ministry. The irrigation season was to begin on June 1. But the only thing the international community and the State Department were interested in was the restoration of the marshes. I had a stream of correspondents and international agency visitors coming to talk with me—all they wanted to hear about was what was being done to restore the marshes. In addition to getting ready for the irrigation season on June 1, we were struggling to patch together, and pay, a workforce of 18,000 people and fix a number of damaged generators that powered the pumping stations that were needed to drive the 3,000,000 ha irrigation system.

The banking system no longer functioned—there were no assets in the banks, and people had to be paid the old-fashioned way: standing in line and getting cash from the back of a truck. The ministry had 18,000 people, scattered in 100 district offices, who managed 10 large dams, 11 major state-owned enterprises (under Saddam, private companies were embedded into each ministry, which helped him launder the Oil for Food money), 10 major barrages (low, large dams across major rivers to raise the water level for distributing irrigation water over very flat terrain), and over 500 pumping stations, all of which were looted and stripped of just about every piece of useful equipment, including huge pumps and generators that were bolted to concrete. I was going to deal with the marshes, but they were low on my list of priorities during that first critical month.

Restoring the Marshes of Mesopotamia (Part 2)

Eugene Z. Stakhiv

"Time to bring back the water!"

When you are in an environment where a certain degree of chaos exists, there are opportunities to exert some initiatives that would not be possible were I working in my home US Corps of Engineers office near Washington, DC. For example, I took on the responsibility of repairing the Baghdad Zoo's water supply and irrigation system, even though it was the responsibility of the municipal department. The animals were dying, and I couldn't allow that to happen. I also encountered and assisted a number of very dedicated NGOs in the field, including Oxfam, Doctors without Borders, groups associated with UNICEF, and other humanitarian organizations. It was an astounding sight for me to encounter these solitary Western aid workers, working all alone without security and help, living in the same primitive conditions as the people they were assisting. These were the true unsung heroes of the reconstruction, and I could only salute their dedication and offer them whatever assistance I could—connect a small pipeline here, drill a village well there, fix a water pump or install a small pump for irrigation. After all, I had a ministry of 18,000 people, most of them idle in the first few months, and many warehouses full of pipes, small water pumps, and other equipment that Saddam had hoarded. We put all these resources to good use, even if it wasn't my responsibility to help these NGOs. They were the ones who were providing the most needed, most immediate assistance to the poor villagers. To this day, I marvel at their courage, selflessness, and dedication, and my greatest satisfaction comes from helping them without having to deal with an onerous bureaucracy that was just gearing up.

My main responsibility (as presented in Part 1), was the restoration of Iraq's complex irrigation system and the marshes of Mesopotamia. To do this, we needed to understand just how much water we could count on from the two main rivers in Iraq, the Tigris and Euphrates. It took us a month to develop a schematic of the entire water management system of Iraq—all the dams, barrages, irrigation canals, and drainage network. This was essential before any modeling could be done. At the same time, we needed to repair much of the irrigation infrastructure before it would function properly. One job needing attention was cleaning the drainage canals and ditches of the vast irrigation system, as they had been neglected and were clogged up during Saddam's time. So why not use people to clean those canals and ditches? I was skeptical of just using people to do this at first, because I saw this as a job for machines (i.e., backhoes, bulldozers, and dredgers) and a logistical nightmare that we couldn't take on; when we could barely pay our own ministry employees, how were we to pay thousands of day laborers? But if we could put local people to work, we could solve two goals simultaneously: first, fix the water problem, and second, get people back to work and earning an income after the unemployment caused by the war.

One of the realities of a situation like Iraq is that everything moves very fast, so you have to be ready to take advantage of opportunities—and there were many more opportunities than we had the capacity to deal with. The opportunity to implement this job-creation program came when it was realized that unemployment and the accompanying unrest was becoming a huge problem. To address this unemployment problem, I was asked by Ambassador Bremer, who was heading the coalition government, to develop a plan for hiring people and identifying the potential irrigation sites where they would work. After presenting my plan to Bremer, I was asked to identify more sites so that we could hire 100,000 daily workers by the end of summer! To many, this goal seemed logistically impossible. How would we pay these day-laborers? I silently agreed that this was impossible, but no one asked for my opinion. Ambassador Bremer turned to me and asked how much it would cost to hire 100,000 per day. I knew that the going wage rate for day laborers was about 3,000 dinars, or about $2, per day, and it would take time to ramp up the program. I quickly calculated in my head and said that $20 million would do for the next three months. Bremer quickly turned to his finance and budget advisor, pointed to him, and said "get Gene the money." It was that quick—that simple—and it took no more than 15 minutes, again, not the way it would have worked back in my home office.

I immediately got my staff to work trying to figure out how to pull off this huge initiative, because Bremer would hold me accountable. The senior ministry leaders were not enthusiastic about this idea, so we had a lot of work to do. I asked them once whether they would question a similar program if Saddam had proposed it, and they responded that this was now a democracy, meaning that they should ponder the issue and vote on the proposal. I responded that given the critical situation that we were under, they would soon recognize that I could be tougher than Saddam, because we were now working to restore services for the Iraqi people - not Saddam's personal and selfish goals. There were many more problems hanging over my head: Mosul Dam was in a dangerous state; gates on some dams were not opening, and the reservoirs were filling up rapidly and dangerously; other gates were stuck in an open position, spilling precious water; the Electric Power Commission wanted us to continue generating hydropower at full capacity through the end of August, while I was worried that we wouldn't have enough water for the irrigation season next year if we drew down the reservoirs. The Ministry of Agriculture wanted to expand the acreage that was to be irrigated by 10%, but it was still difficult to pay salaries to the 100 district offices far from Baghdad, and the upward reporting to all the auditors who were filling the palace in the Green Zone was just beginning to snowball. I needed this jobs program like a hole in the head.

Somehow, by the end of August we had exceeded the target of 100,000 daily workers! This was a truly monumental achievement that Bremer considered as one of his big early successes (as noted in his memoir). That success was due largely to the efforts of one of my Iraqi expatriates, Dr. Hassan Janabi, assigned to my staff in May. As of 2018, Dr. Janabi is the minister of the very same ministry that he helped rebuild. That success greatly improved the image of the ministry in the eyes of Bremer and his staff, and it increased our credibility on other initiatives that we proposed, making it easier to compete for scarce resources that were needed for projects that enabled the immediate restoration of some of the principle marsh areas identified by the ministry planners and engineers. It all started with a simple, small idea born during a field trip to the marshes of Mesopotamia.

Before I left Iraq, I had the chance to visit the marshes. I saw an endless panorama of searing, shimmering heat on a flat, windy, and dusty plain that had once been a green marsh. It was 135°F, with a persistent 20 mph wind blowing dust off the desiccated landscape. For thousands of years, the marshes had been a refuge to where princes and slaves fled from satraps and invaders and hid in the vast, featureless wilderness of tall marsh grasses. Saddam had drained the vast Central Marshes of Iraq to "reclaim" the land for irrigation—this would be his modern paradise. Instead, because of the war with Iran, the projects were never completed, and he left desolation. Following the example of his hero, Stalin, he had relocated the Marsh Arabs, something akin to our Louisiana Bayou Cajuns and Creoles, to other parts of arid Iraq. Many fled to Iran; though Iraq was at war with the Iranians, at least they were Shiites. It was a bleak scene that Saddam created purposefully and relentlessly in southern Iraq, the stronghold of Shiite resistance to his regime.

What I learned on that visit is the dominance of tribal politics—something I did not count on but recognized immediately when we were passed on from tribe to tribe during our trip at the invisible boundaries of their respective domains. It's hard to institute democracy and rights for women, secular groups, and non-Muslims when you have the powerful force of tribal culture and traditions that span millennia. I had thought my job was tough in the ministry; the jobs of the regional coordinators were extremely intricate, with daily frustrations and setbacks, but I came to appreciate the foundational work these people were doing, building democracy from the bottom up, village by village, sheikh by sheikh, individual by individual. I felt that I was a pretty good technocrat, and a good fit for the Ministry of Water Resources, but working at the local level, negotiating daily trivial disputes among irate competing parties, was well beyond my interests or capabilities.

Political freedom often results in chaos. Slogans and good intentions are no substitute for determined governance. The public needs its services. It can't live long on hopes and promises. It takes a great degree of discipline and competence to make governments work, and while poets, writers, entertainers, clerics, and dreamers may occasionally be great leaders, they don't make good managers, which is what we needed to lead ministries when I was in Iraq. As Bremer noted in his memoir, these competent managers were few and far between. So, dear readers, hone those skills. They are important ones to have.

Part III

Adventures in Australia

Building Bridges with Water

Emily Barbour

Deciding on a career path can be an agonizing process. For me, it came down to two main criteria: studying the subject I was most interested in (environmental engineering) at my home university; or settling for civil engineering at a bigger university with manicured lawns and historic buildings. Initially, the latter criteria prevailed, as was the wisdom of an 18-year-old me where aesthetics and moving away from home counted more than the content of the degree.

I applied for a scholarship to study civil engineering. It became clear during the scholarship interview that I did not share the same level of enthusiasm for structures as the professors and fellow applicants. As I struggled to answer questions regarding my favorite bridge, I privately remembered a high school design-and-technology project that involved making a bridge out of matchsticks. I had scored high on aesthetics and rather low on practicality and structural integrity.

This caused me to reconsider my choice, and I enrolled at my home university and began a combined degree in environmental engineering and science. Instead of learning about the structural design of steel bridges, I embarked on a journey about bridges of a different sort—between natural and human systems, between people, and between disciplines. Such bridges have been built through asking questions, learning to listen, and seeking different perspectives.

The journey so far has been an incredible one and has taken me all over the world. Below are some of the key lessons I've learned and the adventures I've had learning them. Water connects people and the planet and will continue to require innovative and interdisciplinary approaches to tackle the immense water-management challenges that face us.

There Is No Such Thing as Natural

One of my first big adventures was as an undergraduate exchange student in Norway. Apart from some initial mishaps (such as courses being taught in Swedish rather than English and my inability to stay upright on icy footpaths), I learned a key lesson about nature. As part of a Norwegian plant ecology course, I embarked on an ecology assignment pertaining to Australian bushfires. I was interested in the impact of controlled burning on vegetation and had been romanticizing about the need to better mimic the natural fire regime. At the end of the assignment, my professor asked me what I meant by "natural." I had completely missed the fundamental fact that there is no single objective state that defines natural. The environment is always changing, whether through processes such as fires, river flows, or interactions with different species, including humans. This is not to say we humans should not reduce our impact on the environment, but it demands a greater appreciation of the complexity of interactions, responses, and adaptations that occur.

This dynamic interaction was evident in the Norwegian heathlands (also present across northwestern Europe), which have coevolved with communities in response to managed disturbances including grazing and burning. These "novel" or seminatural systems are in many cases viewed as having significant ecological value.

This lesson of what is natural was reinforced during my Ph.D. studies where I was interested in trade-offs between delivering water for the environment (in this case wetlands) and for agriculture. Despite setting off with the intention of focusing primarily on optimization methods for evaluating trade-offs, I realized the main challenge was in understanding ecological response and what constituted a desired ecological outcome. It became apparent that what was considered a "better" outcome was a complex mix of ecosystem dynamics and societal preferences.

Learning Through Difference

My second big adventure during my undergraduate degree was walking across campus from the engineering faculty to the psychology faculty (and back again). Rather than selecting more clearly complementary science subjects, I used my combined degree as an opportunity to learn about whatever interested me at the time. This did not come without its challenges, particularly as it meant I frequently did not have the prerequisite introductory courses for a number of subjects. However, I persisted, and even succeeded in convincing the university to recognize additional subjects not previously listed on the options of approved science subjects for the combined degree.

The eclectic mix of elective subjects included microbiology, animal physiology, town planning, animal ecology, and three psychology subjects, not to mention the incredible diversity already offered within my engineering degree including environmental law and a philosophy course on technology and human values. The mix of disciplines helped me see the world in different ways, and it has shaped my career since then. Unfortunately, such diversity is not always appreciated. One day I had a psychology lecture immediately following an engineering lecture. I had just sat in a class where a comment was made by the lecturer about the superiority of engineering problem-solving and quantitative skills compared with the "soft" sciences. Bemused, I headed across to my "soft" science class to hear an equally dismissive comment made about engineers and the lack of appreciation for broader, qualitative contexts.

Although this attitude is definitely changing in the current era of interdisciplinary research, similar rhetoric is still alive and well. We need to move toward a place where we can appreciate and recognize the valuable contributions of multiple disciplines and perspectives.

I have been impressed by endeavors to do just this through two recent projects, one led by the University of Southampton on the Ecosystem Services for Poverty Alleviation (ESPA) Deltas project, and the other by the University of Oxford on REACH—Improving Water Security for the Poor. In both cases, the goal has been to move beyond parallel activities undertaken by different disciplines, institutions, and countries, and instead to work together to collectively improve knowledge and tools for decision making, grow networks, and create impact.

The Power of Listening

A number of years ago, I read a novel called *The Camel Bookmobile* by Masha Hamilton, where an enthusiastic librarian ventures into a remote area of the Kenyan desert to bring literacy and education to a nomadic settlement. Over time, the librarian comes to realize that her ideals on progress and modernization are far from straightforward. The library precipitates conflict and division between long-standing traditions and community cohesion on one side and the younger generation's eagerness to learn about new and far-off worlds on the other. This important message has stayed with me and has played out a number of times in my work and travels. It has taught me the importance of understanding context and listening to different views in order to better anticipate unintended consequences.

Working in coastal Bangladesh through the ESPA Deltas and REACH projects, I became interested in the impact of infrastructure on water security and human well-being. Many communities there live with the risk of extreme flooding, the impact of which is often the most severe on those most poor and vulnerable. In the 1960s and 1970s, a network of embankments was constructed to reduce flood impacts and improve food security. These embankments have been viewed as both a savior and a curse. Initially agricultural productivity improved, but it later declined because of waterlogging. Environmental degradation combined with inadequate infrastructure maintenance has contributed to loss of livelihood, social conflict over land use, and the displacement of households to more vulnerable areas.

One of the most striking examples of this can be seen in remote areas of southwest coastal Bangladesh, where communities live on muddy embankments surrounded by water. Water forms an integral part of their lives and livelihoods. They survive primarily on fish and have to travel by boat to access clean water. These communities are some of the poorest in the country, where embankments have contributed to their displacement but also afford some protection through their elevation. It is crucial that we understand the risks and needs faced by poor and isolated communities in evaluating the effectiveness of interventions, as well as recognize local knowledge and adaptation strategies.

During my Ph.D. research, Australia's Murray-Darling Basin was undergoing a period of significant water-management reform, resulting in tension and conflict over how water should be shared in a water-scarce system. Talking with a diversity of stakeholders was critical to gain insights from different perspectives to better inform the development of quantitative water resources and ecological modeling tools. Government staff, sector specialists, researchers, and community members all had different and important perspectives on the intersection between water, the environment, and society.

To me, these experiences demonstrate some of the incredible opportunities that exist when working in water management, opportunities to learn from diversity, marvel at the exceptional generosity of those who have so little and be moved by the extreme hardship of poor and marginalized people. Although most of us in this field are motivated by making a difference, it is imperative that we continue to listen in order to build the most effective bridges.

Developing Drainage During Droughts

Graeme Dandy

"Modelling drainage during this drought may be difficult."

South Australia is the driest state in the driest inhabited continent on earth. Much of the state has an average annual rainfall of less than 250 mm (<10 in.) per year. However, the Upper Southeast of the state is significantly wetter than this. When the country was settled by Europeans in the nineteenth century, the region consisted of a series of wetlands. To prepare the land for agriculture, the early settlers constructed a series of drains to divert water to the sea and to lower the water table. More recently, drains have been constructed to divert water to the more than 200 significant wetlands in the region and to lower the water table in agricultural land.

Holger Maier and I were involved in developing computer software, called a decision support system (DSS), that could be used to guide the operation of the 65 regulating structures for diverting water to selected wetlands in the region. A postdoctoral fellow (Matt Gibbs) and two Ph.D. students (Daniel Partington and Abby Goodman) were engaged on the project. Matt was to work on the DSS that included surface water models and an optimization component. Daniel worked on modeling the impact of the drains and weirs on surface water–groundwater interactions. Abby evaluated the impact of various water levels and salinities on the ecological health of key wetlands.

The project commenced, unfortunately, during the worst drought on record for Southern and Eastern Australia. The drought lasted 10 years. During the three-year duration of the project, many of the drains had zero flow and most of the wetlands were dry. Developing a hydrological model for the region proved difficult without flow data for the many drains within the region. The data that were available were not representative of current or future conditions because of the extensive construction of drains in the region over more than 100 years. Undaunted, Matt developed conceptual rainfall-runoff and salt-transport models for some of the major catchments in the region. These models formed the basis of the DSS. Abby also had some difficulties as the proposed ecological fieldwork could not be carried out because most of the significant wetlands were dry. Instead she analyzed the dominant vegetation types in the significant wetlands. She also took samples of four of the main species of macrophytes and undertook growth and seed-propagation studies of these species. Trials were carried out for the growth rate and propagation of the plants for various combinations of water depth and salinity as well as for conditions where saline water was supplied at various stages of the growth cycle. The results showed that the plants could tolerate short bursts (of up to six weeks) of saline water provided that this was followed by a period of fresh water.

Daniel developed a fully integrated surface water–groundwater model to identify the contribution of direct runoff, interflow, and groundwater to flow in a drain under various rainfall conditions.

Although the original aims of the project were not fully satisfied, a DSS was produced that has proven to be useful for management of the drainage system. Matt and Abby were subsequently employed by the agency that manages drainage in this area, and Matt undertook further collaborative research on water resources of the Upper Southeast region. Daniel was employed in a postdoctoral position to continue his research. Furthermore, eight journal papers and nine conference papers were produced from the project. Just think of what all of us could have accomplished if we had had any water to drain!

We learned the following valuable lessons from the project:

- It's hard to get support for studying drainage problems during a drought, especially a drought of record.
- The outcome of any research project is not likely to be what was predicted in the research proposal.
- Many useful outcomes can arise from a research study, even if the nature of the study changes significantly from the original aims.

- Even if the real-world impacts are not immediately obvious, one can always publish!
- If research is not challenging, it probably isn't research.
- If research is not fun, it definitely is not worth doing! But when it is fun, it's amazing, and we get paid for doing it.

"Optimizing" a Piped Irrigation System

Graeme Dandy

Laurie Murphy, a Ph.D. student under the supervision of Angus Simpson and me at the University of Adelaide in Australia, was the first person to apply the technique of genetic algorithms (GAs; a randomized search procedure) to the optimization of water distribution systems. Many researchers had tackled this problem with varying degrees of success since around 1969. GAs showed considerable potential to crack this difficult combinatorial optimization problem.

Before Laurie completed his Ph.D., the three of us started to undertake consulting projects applying GAs to optimize the design of actual water distribution systems. Our timing turned out to be good, as a number of irrigation authorities in Australia were in the process of replacing channels with pipes for distributing irrigation water. Channels suffer the disadvantage of having significant losses through seepage, evaporation, and spillage; whereas, pipes have much smaller losses. It is also more difficult to control water levels and flows in channels, particularly when irrigators can start or stop extracting water at unexpected times. Consequently, there is usually a requirement for irrigators to order water in advance, so that releases to the channels can be programmed to minimize fluctuations in flow and water level. Pipes allow water to be extracted by any irrigators at any time, provided the capacity of the distribution system is not exceeded. Two of our first consultancies were undertaken for irrigation authorities who were planning new piped distribution systems to replace channels. One of these was for an irrigation authority in the state of New South Wales (Australia). This was a two-day drive away from our homes in Adelaide.

The operation of the proposed scheme involved pumping water out of the Murrumbidgee River into a large holding tank on a hill and allowing the water to flow under gravity through the piped distribution system to the various irrigators. Our job was to size the pipes so that a set of peak demand patterns could be satisfied throughout the scheme. Before travelling to the site, we had carried out some preliminary optimization runs and found that considerable cost savings could be achieved by raising the maximum water level in the proposed new tank by one to two meters. When on site, we discussed this with the engineer from the irrigation authority. He dismissed it fairly quickly by referring to "aesthetic considerations" that limited the maximum acceptable height of the tank. After travelling around the site, the engineer took us to the site of the proposed tank. As we walked up the hill, the engineer referred to a house nearby. "That's my place," he said proudly. We immediately realized the importance of the "aesthetic considerations" to a key decisionmaker.

Any optimization study needs to take into account the constraints on the system. These can be physical, technical, legal, financial, or could represent the values of key decisionmakers as expressed through "aesthetic" or other values. We completed an optimized design for the system that satisfied all of the constraints (including the aesthetic ones) and managed to save around 15% of the capital costs of the project compared to its original design.

Part IV

Adventures in Europe

Years of Experience versus Computer Models

Thomas Ackermann

The River Mosel is an international waterway connecting France and Luxembourg to the River Rhine in Germany. It forms a common border between Luxembourg and Germany until it enters the Mosel valley in Germany near the Roman city of Trier. The reservoir Trier is the first in a cascade of ten reservoirs in the German portion of the river, each equipped with locks and hydropower stations. The allowable storage level variation in each of these reservoirs is limited by the navigation authority to ± 5 cm.

As a young scientist working at the Aachen University of Technology, Germany, I was given the task to set up a team to develop and implement a real-time decision support system (DSS) for operating the reservoir Trier.

Together with two research assistants and with the support of the electrical engineers in our hydraulic laboratory, we spent about nine months on the project. Among others, Pete Loucks from Cornell University advised us on the importance of getting the operating staff on site engaged in the design and calibration of our models of the system.

Part of our challenge was dealing with the unpredictable inflow into the reservoir Trier. Flow variations of 50% compared to the mean discharge over a time period of 10 minutes at random times was common. These flow variations were caused mainly by the operation of upstream reservoirs in France, including the operation of ship locks, which caused large volumes of water to be released within short time periods. We were told that past attempts to work with the French authorities to attenuate these waves had not been successful.

To control flow throughout the German cascade of river reservoirs and thereby to ensure the tight water level tolerances, these flow variations had to be attenuated in the reservoir Trier. To complicate the situation, the Rivers Saar and Sauer also flow into the reservoir. In addition, navigability of the River Saar also required the operation of locks. To meet the downstream flow tolerances, the allowable storage level variation in the reservoir Trier, and only for this reservoir, was raised to ± 15 cm (i.e., to a range of 30 cm).

The idea behind our real-time DSS for operating the reservoir within these tolerances was to calculate an optimal trajectory of future release decisions at the hydropower station as well as the associated water level elevations. From this trajectory, only the first release decision would actually be used to operate the power station. Five minutes later, the next release decision would then be taken from an updated trajectory, and so on.

To validate the model, we had to measure flows at several cross sections along the reservoir. We did this using a measuring device mounted on a small rubber dinghy, which was way too small for the device and the three of us forming the crew. Luckily, we did not capsize when the large container ships crossed our path. Surely, it was quite dangerous as we tried to surf the waves as they came.

A major constraint in our model was to ensure that downstream water levels did not rise above an upper limit that would restrict navigation under bridges at all times. Water levels also had to not fall below a lower limit to guarantee the minimum water depth. At the same time, the release could not exceed the capacity of the turbines to avoid spill and an accompanying loss of energy. In addition, the amplitude of water level change could not exceed the maximum amplitude that had occurred during the previous years of manual operation. Finally, the reservoir release flow rates were to be held constant as long as possible to reduce the wear and tear of the mechanical equipment.

By listening to the operators, we learned a lot about the hydrodynamics of the reservoir Trier and the entire cascade. Many of the operators had spent their entire professional career at the control station. We observed that each of them followed a slightly different individual strategy on how to operate the reservoir, and each of them had good arguments justifying why they did what they did. They were very good at anticipating the unnatural inflow variations and their impact on water levels in the reservoir Trier. We also witnessed that they adjusted their forecasts as soon as updated information became available. Their intrinsic approach on how to operate the reservoir was very similar to the one we proposed to incorporate into our DSS model.

When we introduced our modeling approach, the older, more experienced operators tended to be critical toward the new control policy. They feared competition from the computer-based system. But, most importantly and most honorably, they took real ownership and were sincerely concerned that our model algorithms could not cope with the complex inflow conditions and restrictive water level tolerances.

During the first test of the prototype on site at the central control station, the staff on duty frequently interfered manually when they believed that the system did not perform as they had expected. The test therefore did not deliver sufficient data to further develop and calibrate the simulation and optimization models. A breakthrough in the model testing was eventually achieved when the responsible managers of the utility company clearly demonstrated their support and willingness to carry out the test. They agreed to test our model over a limited time period of three days. During this time, the operating personnel on duty were released from all responsibilities and liabilities for the operation. All parties involved were present over the entire period of the test. Any changes to the proposed operation would then be discussed and implemented in agreement with everybody attending the test. During this period, the new system performed without fault. No manual interference by the staff was required. However, the flow was unusually steady, so the performance could not be considered representative.

Still, the trust in the system began to rise. It was agreed to extend the test by seven days. From now on, only the staff would be present at the control station. The operators would only interfere manually if the water level tolerances would otherwise be violated. In addition, they were asked to document any changes in release that they would have made, but for the purpose of the test they were to refrain from actually making the changes.

Again, most of the interferences and comments from the staff were based on a sound and intuitive understanding of the hydrodynamics. While discussing the findings with the staff, we could pinpoint and analyze those instances where the manual interference was indeed required to avoid a violation of the tolerances. We could also highlight that some of the proposed changes had actually not been necessary. With this information at hand, the system could be developed further, and the staff increasingly gained confidence in the tool. The younger employees especially were more and more keen on the system being implemented. Unlike their more experienced colleagues, they did not have the knowledge about the hydrodynamics, and they expected that the system would provide them valuable support in operating the reservoir within the tighter tolerances.

To be present on site and to take part in the day-to-day operations was clearly essential for the development of the system. In addition, it was highly important that the representatives of the parties involved had the power, authority, and willingness to provide the necessary data. Some of these data came from new monitoring facilities that they installed and paid for just for our benefit. The management team of the utility company proved to be courageous in the trust they gave us. Right from the beginning, it was clear to all that any malfunction of the program could cause a major violation of water level limits downstream in the cascade, thereby impacting navigation. A reliable operation of the hydropower plants is fundamental for the utility to keep its license to operate.

The decision support system was eventually implemented as a key component of the overall control system for the River Mosel. We enjoyed being involved in the creation and implementation of our DSS. We learned that building up confidence, while taking time, is crucial for an effective implementation of such systems. The project brought us to one of the most picturesque areas in Germany, the Mosel valley. We spent a lot of time on-site, not only engaging with the local staff and calibrating our model but also testing all the great food provided by local butchers and bakeries around the area. By the time we sampled and resampled their many delicious products, we considered writing a food guide if only we had had more time. Our involvement in the operation of the reservoir Trier was too short. What a pity.

Technical and Cultural Exchanges in Ukraine

Lily Baldwin

"You wouldn't believe how many international engineering events end with karaoke."

As the impacts of climate change are being realized in our world today, there is renewed interest in non-carbon-based energy sources, including nuclear energy. I don't think we have the luxury of writing off any option at this point. But let's not forget the potential and real dangers, and the difficult lessons we learned from previous experiences. In reminding myself of the accident that occurred in Chernobyl in April 1986, I was a bit disconcerted by how whitewashed accounts are on the Internet. This story is about my own experience visiting the Chernobyl Exclusion Zone in 2002 and what I saw.

After the Union of Soviet Socialist Republics (USSR or Soviet Union) formally dissolved in the 1990s, former weapons scientists needed alternative applications for their research. I was working at Lawrence Livermore National Laboratories (LLNL) in the first decade of the 21st century, and the Department of Energy (DOE) was encouraging partnerships between environmental programs at the national laboratories with former Soviet Union weapons scientists. To this end, a handful of us from LLNL were selected to participate in a conference in Ukraine in 2002 with researchers from several of the former Soviet Union states. I presented on sediment movement in urban stormwater runoff and the policy and regulatory framework of stormwater regulations in the United States. The conference was held in what is called a sanitarium, which does not have a precise English word translation. Loosely, a sanitarium was a place where people went to relax and become rejuvenated. The sanitarium where our conference was located was just outside of Kiev.

In preparation for this trip, I had learned the phonetic sounds of the Cyrillic characters used in Ukraine. Many of the signs were English words spelled out in Cyrillic characters. For example, I found a sign at the hotel, that when sounded out, phonetically spelled "Business Center." The conference was three days long, and everyone was very reserved until the conference banquet. As one would expect in a former Soviet Union state, this banquet included a lot of vodka—a bottle every few place settings along the long rows of tables. There was a lot of toasting, story-telling, and thank you speeches. Eventually, it was the US delegation's turn, and I was tasked to sing a song. I chose a song I felt would be a familiar melody from the *Sound of Music* musical. But in keeping with the vodka, I chose the lyrics from a rugby drinking song. This sparked an international sing-off with songs in Russian, Ukrainian, Uzbek, and Kazakh. The singing continued until the Georgians decided to recite a romantic poem instead of singing. And the vodka had run out.

After the conference ended, our LLNL delegation moved to a hotel in the heart of Kiev, and I had the unique opportunity to take a day trip to the Chernobyl Exclusion Zone. When I was first asked if I wanted to go, my immediate reaction was, "heck, yeah." But as the magnitude of the accident at Chernobyl finally hit me, I reached out to one of LLNL's many industrial safety subject matter experts and asked her if I should be concerned about having babies if I went. She advised me to take advantage of this once-in-a-lifetime opportunity. She said that one day in the exclusion zone should not be a problem to my reproductive system but to only bring home memories and nothing else. She also asked if I would be willing to wear a personal radiation monitor that records instantaneous radiation measurements. The personal monitor is unlike the dosimeters that everyone who worked at LLNL wore, which measured only cumulative exposure. Unfortunately, I was unable to take home any data; every time I turned the personal monitor on, it kept beeping its high-level alarm, so I stopped trying.

The primary purpose for the Chernobyl Exclusion Zone field trip was to review two research projects on environmental remediation technologies being conducted within the exclusion zone. Our hosts picked us up from our Kiev hotel bright and early, as it would take several hours to drive there from Kiev. Our first stop was to the abandoned administration buildings, which included a lobby akin to a hotel. There was a diorama of the surrounding area that gave a good perspective of the area, including reactors' location relative to the river. The Dnieper River provided the cooling water for the reactors. We then headed to the first research project, with its combustion research lab housed in warehouses about a mile away from the Dnieper. Driving along the dirt road to the warehouses, we saw boats of all sizes, permanently positioned where they landed after the explosion blew them out of the river. The relatively small combustion experiment was anticlimactic after that.

Our next stop was to the gatehouse of the wall surrounding the reactors that had been converted to a form of visitor area. We were shown a video of what happened the day of the accident, and the ensuing emergency response and evacuation efforts of about 160,000 people living in the surrounding vicinity. These residents were evacuated over a two-day period, predominantly from the city of Pripyat where most of the power-plant workers and their families lived. The video highlighted the firefighters who continued to fight the blaze even after they had exceeded their exposure limit given the radiation levels immediately after the explosion. These firefighters are the unknown heroes who saved thousands of people (one could argue the world!) by containing the fire as quickly as possible and preventing an unimaginable disaster. Most of these firefighters died within a few days of the explosion from acute radiation exposure. Even in 2002, 16 years after the explosion, my colleagues from LLNL were bewildered at ambient radiation at the gatehouse—radiation levels 300 times the background level at my home in California. The solution to the April 1986 accident was to pour a lot of concrete over the fire. But it was not 100% coverage. Over the decades, animals have traveled in and out of the "contained" reactor area. Few people probably know that a second reactor, separated only by a wall from the reactor near where the accident occurred, continued to operate and provide power to the thousands of people it served until it was finally decommissioned around 2000.

The name *exclusion zone* is a misnomer. Although it is technically illegal to live within its boundaries, at the time, there were about 200 researchers living there under exemptions to the law. We had lunch in the small town of Chernobyl and were fed by some of the warmest, kindest people I had ever met. I didn't have the heart to ask from where the food I was eating had come. But I was most affected by the life-size statues of the firefighters fighting the fires after the explosion. The statues were not very tall, all near my own height. The artist did a beautiful job of capturing the expressions of the men who carried unimaginable responsibility as they struggled to keep the fire from exploding the nuclear reactor. Can you imagine what our world would have been like today if that nuclear reactor had exploded? But for the strength and courage of these men, we will never know. It still saddens me that only a handful of people will ever get to see these statues of giants of men.

We visited the second research experiment after lunch. Again, the laboratory tour itself was overshadowed by the drive there. As we passed miles and miles of former farmland, every once in a while, I'd see an elderly gentleman working his land. According to our hosts, in addition to the 200 or so scientists with exemptions to live in the exclusion zones, there were some former residents who returned after the evacuation and refused to leave their homes.

It was late when we finally left, and the guard at the gate waved us goodbye. He did not bother to wave a Geiger counter wand at the bottom of our shoes. The readings would have been too high. I left my "hot" shoes at the hotel rather than take them home.

In summary, I learned

- How music transcends language barriers;
- An appreciation for good vodka;
- That boats aren't meant to fly; and
- A handful of firefighters saved the world after the accident at Chernobyl.

Long-Term Water Resources Planning—Almost 50 Years Later

Janusz Kindler

"This project not only helped Poland, but it also was a great experience for all of us."

The control and use of water has a long history in Poland. Compared with many other European countries, nature has bestowed on Poland limited water resources and great variability of their occurrence. Initially, water resource managers were concerned predominantly with flood control, river navigation, and hydroelectric power production. The period following World War II, however, saw a distinct shift toward multiple-purpose water projects designed to meet the needs of the population and economy, as well as to restore and develop agriculture, industry, and urban centers devastated by the war.

In 1956, I graduated from the Warsaw University of Technology, Faculty of Hydro-Engineering. My interest in water goes back to the books on brave men building dams in the wild forests—great adventures in the mind of a teenage boy! Following graduation, I got a job in Hydroprojekt, the largest engineering organization in Poland, which designed most of the large water structures in Poland. This was a great post-graduate learning experience. Hydroprojekt had a group of highly experienced engineers who took very good care of us "green" engineers, as we say in Polish.

In the early 1960s, Poland signed a treaty with Iraq, taking over a large Amarah development project involving reclamation of land areas of saline soils in the Lower Tigris valley (250 km^2 about 300 km south of Baghdad). Hydroprojekt was one of the few Polish firms that could undertake such a large project, and all of a sudden, I was invited to join the Amarah project team as an assistant chief engineer. At that time, I already knew the English language fairly well, and this probably was my greatest advantage. The Amarah project was my first professional contact with an arid climate—two years of most interesting tasks on irrigation and drainage, hydrology, and water management in climatic conditions very different from those at home.

In 1968, Hydroprojekt got involved in preparations for a new water resources development project. This was to be a comprehensive study strengthened by the assistance of the UN/UNDP Special Fund, under the name "Planning Comprehensive Development of the Vistula River System," in short, the Vistula River Project. The Vistula River is a major river flowing through much of Poland. The goal of the project was to formulate a water resources investment program capable of meeting multiple water demands and uses in the Vistula Basin, looking forward to the years 1985 and 2000.

The UN/UNDP Special Fund allowed us to invite foreign experts for consultation and advice concerning different aspects of the project. To my surprise, as I had no political (Communist Party) credentials or connections I was invited to be acting project co-manager and hold this title for the first half of the project.

One of our first tasks was to select a consulting firm whose experts could work directly with us, especially on mathematical modeling issues. The person recommended by UN/UNDP to go to Warsaw and to write a draft request for proposals from different firms was Dr. Daniel (Pete) Loucks, a young professor from Cornell University in the United States. This was my first encounter with Pete almost 50 years ago! The consulting firm that won the contract was Water Resources Engineers, Inc. from Walnut Creek, California, headed by internationally well-known professor Gerald T. Orlob. Jerry. Other experts from that firm also spent considerable time with us in Warsaw, and 10 of us Poles had the chance to spend three months of training in Walnut Creek. Such exchanges are no big deal today, but at that time it was a unique opportunity for us Eastern Europeans to visit the West, and indeed the United States.

There were two principal Polish institutions directly involved in the Vistula Project. The first one was the Hydroprojekt. The second institution was the Institute of Environmental Engineering of the Warsaw University of Technology (IEE WUT). Participants from the university were already familiar with the application of new system analysis methods in the area of water management (I joined IEE WUT after completion of the Vistula Project). The director of the Institute was Prof. Zdzislaw Kaczmarek. He was internationally known for his work in hydrology and statistics.

A starting point for our studies was to spatially decompose the Vistula system into a series of interdependent components. Such decomposition was needed because of the exceptional size of the Vistula River basin, the large number of water uses, complicated system structure, and the limited computer facilities available at that time in Poland. We also decomposed the management problem into the purposes water was to serve. This included water supply for the population, agriculture, and industry; pollution control and maintenance of the minimum acceptable flows; flood control; hydropower production; and development of recreational facilities.

I recall spending a lot of my time helping develop the so-called Water Resources Management (WRM) Model. Once each investment variant was defined for a given subsystem, we evaluated its performance using a simulation-optimization procedure (called the Three-Step-Method). Our goal was to find water allocation policies that would meet target demands. When an allocation fell short of the target, it was defined as a deficit allocation. Each deficit was weighted by a penalty factor reflecting the relative priority of each water use over all other users. At the beginning of the study, the intention was to estimate these penalty factors in monetary terms, but we soon learned that some deficits could not be expressed strictly this way.

The Three-Step-Method was a product of joint efforts by a mathematical modeling group comprising experts from Poland and from Water Resources Engineers, Inc. The method involved three computer programs, which were applied sequentially in order to (1) determine a set of target releases for individual reservoirs in the system, (2) develop operating rules for the reservoirs given the inflow hydrology and the target outflows, and (3) determine the optimal allocation of available water to all water uses considered in the model, given the operating rules from (2). This method was made operational on the Polish-made Odra 1204 and 1304 computers, but it was a challenge because of the limited capacity of these computers. Later, the method was transferred to computers in Walnut Creek, and our attention shifted to the development of the so called Single-Step-Method.

The Single-Step Method used a solution procedure appropriate for solving network flow problems. The physical system was represented by a node-link network that included river reaches, demands, supplies, return flows, reservoirs, and junctions. Since the network is only a spatial representation of the problem, it had to be expanded to include time periods as well. This was accomplished by introducing the storage components, in which the final storage volume from one period becomes the initial storage volume for the next period and so on.

The next part of the project was restricted to investment alternatives, which survived the screening process done with the help of the WRM Model. Finally, two alternative programs were formulated for the target year of 1985, two for 2000, and two for ultimate development of the basin's water resources. The programs were arranged this way so that the highest reviewing authority, the Planning Commission of the Council of Ministers, could evaluate the model results, including their economic and social impacts.

To conclude this story, I need to admit that while the results of the UN/UNDP Vistula Project were praised and approved, their implementation was rather limited because of limited state financial resources. Some elements of the investment program developed in this study are still under consideration and being debated. But for many Poles involved in this project, it was the first, and very useful, attempt to work on an international team. Following this project, some of us became leaders at international research organizations, and others joined engineering faculties in Poland and abroad. We were all preaching the use of systems analysis approaches for identifying, analyzing, and evaluating alternative solutions to water resources management problems. Memories of this work done almost 50 years ago with participation of international experts from both sides of the iron curtain are never to be forgotten. The UN/UNDP-supported Vistula Project made a huge impact on the professional development of many young Polish engineers. In addition, many of the friendships developed in those politically challenging years are still strong and give us pleasure today.

Working Behind the Curtain

Pete Loucks

Early in my career, I spent some time in Poland helping prepare a proposal to the United Nations to fund a study of the management of the Vistula River and the development of its basin (See previous story by Janusz Kindler.) The Vistula is a major river in Poland that flows through the country, including the capital city of Warsaw. The proposal involved the use of "modern" systems analysis methods that included simulation. The idea of simulating was meeting some resistance in the Political Bureau (Politburo) of the Communist Party in Poland. To overcome this resistance, we had to convince them that this kind of simulation differed from the type normally associated with that word in Polish. At that time, "simulation" referred to those who were pretending (simulating) to be in poor health in order to be excused from work.

The project was funded, and a plan was developed based in part on simulations performed on the biggest, but still slow, Polish Odra computers of that time. In my opinion, the real benefit of that adventure was the training of many bright, young Polish engineers and scientists, who later became leaders in their fields, earning international reputations. (See the story by Janusz Kindler.) I am still in touch with many of them. I observed that Poles enjoyed telling jokes on themselves and their political situation and knew how to consume hot chocolate spiked with vodka warmed by a bonfire in the middle of a snow-bound forest on a freezing night. They were free and having fun, compared to those whom I met in Romania a few years later.

My introduction to Romania, also early in my career, involved being part of a group asked by the United Nations to apply systems analysis methods to the Upper Mures River Basin in the northern part of that country. This basin had just experienced flooding and earthquakes and was in need of assistance in its redevelopment. I found it much easier to enter and exit Romania than Poland, but once inside the country, the restrictions placed on the people we met were much more evident. If we had private conversations with anyone, they had to report the content of those conversations to security officials. If we gave cigarettes to anyone, they had to give them to their superiors. All of us were being observed all the time.

I recall the first meeting of our group, sitting around a bare table, discussing with our Romanian colleagues just what development options we could, and should, consider. When we asked if we could look at a map of the river basin we were to study, they said no. I asked if I could buy a road map of the area. Again, no. I'm not sure what we accomplished at that first meeting, but it motivated me to bring to our second meeting a large pile of aviation maps that I had access to back home in the United States. Those maps had considerable detail on them that could be useful to our study. As soon as I put them on the table, they were taken away by our observers. I asked them how many more they would like, and from whom were they keeping them secret? Certainly not us. Eventually, they allowed us to work with maps. By the time the project was over, all but one of our Romanian counterparts had left their country. As beautiful as Romania was, and is, it did not seem to be a place where people were having fun, as subsequent history showed.

Changing a Culture in Portugal

Pete Loucks

"No more dams until we develop an integrated plan."

With support from NATO, Dr. Luis Veiga da Cunha, then a division director of the National Laboratory of Civil Engineering in Lisbon, initiated what turned out to be a 10-year project that changed the culture of water resources planning and management in Portugal. Prior to the conclusion of that project, the government agency responsible for water resources was more like a public works department interested in and very capable of building dams, whether needed or not. They were experts at that. Dr. da Cunha's vision, as I viewed it, was to change that culture to one of more integrated planning that took into account demands as well as supplies, benefits as well as costs, and environmental and social impacts. I was invited to contribute to that effort, one involving an amazing team of young engineers and scientists. Some members of this group became politicians themselves, such as the mayor of Lisbon and the Secretary of State of the Environment.

Initially the existing water engineering establishment was not at all supportive of this project. They assigned the project team to a highly polluted river in the north of the country to test the team's ability to clean it up. The politicians representing the Communist Party in Portugal turned down the team's request to work in the south where people go for vacations. They told us they did not want to do anything in that region that might improve the conditions of many workers living there who supported the Communist Party, for fear that those workers living in better conditions might then support another political party (can you believe that!). The river the team was allowed to work on could change its color daily depending on the dyes used by the textile industries that discharged their wastes into it.

Working very closely with all stakeholders, including the polluters, the team eventually was able to establish a river basin water quality management plan acceptable to all, but more important, they were able to establish an institution that had the authority to regulate pollutant discharges and implement integrated water resources planning in the basin. Progress accomplishing all this over a 10-year period was not always smooth and positive, as political parties and their enthusiasm for this effort changed, but it worked. Some of the credit may be because of the innovative, interactive graphics-based modeling we did that made communicating with and involving stakeholders more effective. But clearly, the majority of the credit goes to the leadership of that project and the level of excitement in the team that came from knowing they were doing something big that was being noticed by the news media and, indeed, the public.

Insiders, Outsiders, Funds, and Perseverance: Ingredients for Transboundary Water Management

H. P. Nachtnebel

This story describes the development of a transboundary water management plan for the Danube basin emphasizing the role of external experts, funding, and nongovernmental organizations (NGOs) over several decades. To me, it is a good illustration of how politics and governance mix with science in the real world of water management. As a scientist, I observed and participated in much of what follows.

The Danube Basin in which I live is one of the most international river basins in the world. The basin includes areas in Germany, Austria, Slovenia, Czech Republic, Slovakia, Hungary, Croatia, Bosnia-Herzegovina, Serbia, Montenegro, Kosovo, Ukraine, Bulgaria, Romania, and Moldova. Smaller watershed catchments in Switzerland, Italy, Poland, and Albania also drain into the Danube Basin.

Until the late 1980s, water-related transboundary political issues in the Danube Basin were addressed in official bilateral ministerial-level meetings. Basin-wide scientific perspectives were mostly addressed at scientific conferences. Later, these regional conferences were organized as part of the International Hydrological Program (IHP) of UNESCO.

Addressing upstream-downstream problems in the Danube, such as water allocations and pollution, calls for a basin-wide approach. This more comprehensive approach was mainly promoted by Romanian officials, who felt that most of the pollutants originated in the upstream countries. Political pressure was building in Romania to do something about it.

One way riparian countries could achieve closer cooperation was to agree on the degree of pollution that existed in the Danube. At that time, each country had its own data, which led to contradicting national conclusions about the environmental state in the basin. An agreement was reached in 1986 promoting the implementation of a basinwide standardized water quality monitoring system and the sharing of data among riparian partners. Before the full implementation of the monitoring system could be achieved, the political system of the former communist countries in Eastern Europe, including Bulgaria, Czechoslovakia, Hungary, and Romania, collapsed.

In conjunction with political changes around 1990, new external players entered the scene, and new initiatives were launched to assess the environmental state of the Danube and to establish a basin-wide environmental management strategy. The first coordinated transnational survey along the whole Danube was conducted by the Cousteau Foundation in 1991 to1992. They monitored the entire main stem of the Danube by using their own ship-based laboratory and helicopter, accompanied by press meetings in all major cities along the Danube. This activity definitely contributed to an increased public awareness of environmental problems.

In July 1991, UNDP launched a program titled "Environmental Management in the Danube River Basin." Its purpose was to find a long-term solution to the pollution problems in the Danube River Basin. The riparian countries, together with interested international donors and NGOs, met in Sofia in September 1991, where they agreed to develop and implement an Environmental Programme for the Danube River Basin. To oversee and carry out this task, they formed a Danube task force. In addition, a Programme Co-ordination Unit (PCU) was established in February 1992, mainly funded by European Union resources, to implement and manage these activities.

The Gland, Switzerland, office of the International Union for Conservation of Nature was asked by the PCU to develop a coherent overview of the environmental state of the Danube basin. On the basis of a detailed questionnaire, the Danube basin countries prepared national reports that were then reviewed by an expert group. I was a member of that group. Our task was to analyze all the reports emerging from studies under the program and to synthesize them to achieve an agreed-on description of the environmental state in the entire Danube basin. To achieve this task, we organized and held meetings with ministries in riparian countries to resolve discrepancies in data and statements. Some of the meetings with national experts were rather cumbersome, as we were accused of criticizing some of the contents of national reports and not acknowledging some of their national environmental preservation achievements. Some data contained in national reports were even questioned by experts from those same countries.

Finally, after a workshop held in 1994 in Budapest with all country representatives and NGOs, our report on the current state of the environment in the Danube River Basin was approved by the Danube task force. It included a ranking of pollution hot spots in the basin and proposed measures to clean them up.

Our report became part of the first draft of the Strategic Action Plan (SAP), 1995 to 2005. The Strategic Action Plan included input from the PCU, the World Bank, and the United Nations Development Programme (UNDP), as well as from four representatives from riparian countries including Romania, Bulgaria, Hungary, and Austria. Many meetings and workshops were held giving opportunities for representatives from relevant ministries and municipalities, research institutions and laboratories, private sector enterprises, interested NGOs, and environmental journalists to discuss and comment on the report. Results of these consultation meetings were used to revise the SAP report. Finally, after the draft SAP report was completed, a second phase of the consultative process at the task force level was held to achieve full agreement among the countries.

This summarizes how a basic consensus document describing the environmental state in the Danube Basin was prepared and agreed on among the Danube countries, as well as the development and implementation of the strategic action plan for the Danube. The process has been mainly driven by individuals from universities and NGOs, whereas the decisive power always has been with the Danube task force composed of national governmental representatives. Although the external experts were not immune to being criticized, they were considered unbiased and thus trusted to have no national interests. Their conclusions were approved by the task force. Also, external funding helped substantially to achieve an agreement among riparian states.

The PCU has since been replaced by the International Commission for the Protection of the Danube River, located in Vienna. SAP has been revised twice, and major actions have been executed, resulting in an improved environment. Numerous joint projects have been launched to cope with various risks, such as floods and accidental spills. Substantial progress in transboundary water management has been achieved within the framework of the Danube River Protection Convention. Involvement of external experts and NGOs, together with seed money from outside, has supported substantially this basinwide approach.

The Danube story can be considered a success story. But it illustrates that involving, informing, debating, and negotiating with all stakeholders is neither cheap nor quick, and it requires persistence. The hope, however, is that once any consensus plan, policy, or strategy is reached, more sustainable and long-lasting results will follow.

A Conversation During an Athenian Drought

David Purkey

"The wells are drying up... something needs to be done."

I began my undergraduate education pursuing my keen interest in the humanities, in particular philosophy. My interest in hydrology came later when I took an introductory geology class. Once I switched my major, I packed those philosophical books away and began my career in the field of water resources management. It has been a rich career working on the development and deployment of water resources modeling tools to support water managers in their efforts to improve conditions in watershed and river basins around the world. In these efforts, however, I occasionally recalled those old texts, as I struggled to understand the proper place for our increasingly sophisticated analytical tools within what are essentially social and political decision-making processes. Recently, these recollections inspired me to dig back into some of those texts to see what insights they might offer our profession in the 21st century.

Everyone talks about the weather. We have and we always will. Thus, it is not impossible to imagine that two leading minds of the Western philosophical tradition would have discussed the drought conditions affecting Athens around 360 BCE. At that time, Plato, who was 67 years old and the leader of his own academy, and Aristotle, who was 24 years old and studying with him, were engaged in a dialog about the proper role of the state (government) during a drought—and coming to somewhat different conclusions. Looking back on this dialog from the perspective of the Renaissance, Raphael, in his fresco titled *School of Athens* depicted Plato gesturing toward the sky, where presumably some divine truth about society could be discovered, while Aristotle gestures toward the ground where mortals walk. Assuming Rafael captured them discussing the Athenian drought of 360 BCE, what might we expect them to be saying to one another?

> **Plato:** Things are getting worse. The wells are drying up, crops are failing, and conflicts over access to the cisterns still holding water are increasing. Something needs to be done.
>
> **Aristotle:** Might I suspect that you favor some bold, ambitious plan?
>
> **Plato:** Of course. You recall how Peisistratos responded to earlier water shortages by constructing his aqueduct that filled the fountains of the Acropolis. Why are we not considering similar responses today?
>
> **Aristotle:** Because he was only able to achieve such a system by imposing severe taxes on his subjects. Surely you are not proposing that we sacrifice our Athenian democracy simply to increase water supplies. You must know that the empowered citizens of today's Athens are loath to fund such autocratic endeavors.

Plato: But such ambitious responses need not be autocratic. Even though the citizens are fixed in the shadows, so to speak, ignorant of the real options to manage current water shortages, I am confident that someone driven by the quest for the pure drought-management option could identify the correct response that would truly benefit the polis.

Aristotle: It sounds like you would consider the role of officer of the fountains to be one fit for your ideal philosopher–king.

Plato: I would, and that seems to be consistent with the way our democracy is heading. I have seen your notes for your pending treatise on the Athenian democracy, where you document how the position of officer of the fountains is no longer filled by lot.

Aristotle: Indeed, but as much as we cannot let a small group of recalcitrant citizens, based on self-interest alone, reject any collective action to respond to the drought, we must also avoid the temptation to cede all authority to a small group of the elite. You know my views on the importance of a strong middle class. That group must ultimately arbitrate the merits of any response to the current drought.

Plato: That sounds to me like a recipe for paralysis. Surely something as important as the management of water during these dry times warrants ceding control to those with the knowledge and experience to discover the correct course of action.

Aristotle: I am not sure that the officer of the fountains alone could identify such a correct course of action when each citizen brings to the decision his or her own desires and passions. Better to engage the polis in the process of deliberating on what to do. Such a process might not yield the universally perfect drought response, but it might yield something that could practically be implemented. In addition, there is the issue of uncertainty surrounding this decision. How do we know if the drought will persist or the rains return?

Plato: Uncertainty is indeed tricky. Although some would accuse me of being an absolutist in my pursuit for the truth, be it in terms of the proper drought response or any other issue confronting the polis, I am cognizant of my unfortunate friend Socrates who recognized the wisdom of acknowledging what he did not know.

Aristotle: It seems then that perhaps you should be willing to ascribe similar wisdom regarding the unknown and uncertain to the officer of the fountains. Adhering to an absolute version of the true drought response is not reasonable. Can you defer to the deliberations of the middle class?

Plato: I cannot. The issue of how to navigate the current drought is simply too important to the survival of the polis to be left to the middle class. I adhere to my conviction that in this case the officer of the fountain must indeed be granted the authority of a philosopher–king. Given the gravity of the situation and import of our response, it would seem that we are not of a like mind as to how best to respond.

Aristotle: So it seems. Let us hope it rains soon, and if it does not, let us hope that our political deliberation, in whatever from it takes, leads to a decision. Doing nothing does not seem to be a viable option.

History records that in 346 BCE, and again in 333 BCE, Kephisodoros of Hagnous and Pytheas of Alopeke, respectively, were named in decrees commending their efforts as officers of the fountains to restore the public water system of Athens. Archeological evidence also points to increased investment in public water supply in the third quarter of the Fourth Century BCE. Therefore, it seems that the citizens of Athens did take collective action in response to the drought of 360 BCE. We can only imagine the conversations between the officers of the fountains and the citizens of Athens that took place in the Agora, in more than a decade between 360 and 346 BCE, leading to this apparent change in water policy. One can surmise, however, that they may have been contentious.

Implications for Water Management Today

This hypothetical dialogue is offered in order to suggest that, as in antiquity, the current efforts by water management professionals to convince, through analysis, the wider body politic that a particular course of action is in its collective best interest is rooted in a longstanding philosophical debate. Thanks to the early pioneers in the field of water modeling, today's water management professionals have access to powerful tools as they seek to discover the best water management option in the face of uncertainty related to climate change, demography, economic development, and regulatory reform. New approaches to decision-making under uncertainty based on rigorous analysis have been proposed to assist in this effort. However, the fundamental question of whether the "perfect" option, the Platonic form, even exists to be discovered remains, motivating new questions as to the proper role of data and analysis in the creation of knowledge that will guide society towards more sustainable patterns of water use and conservation as pressure on this vital resource grows.

Perhaps the proper response to this question lies not in rejecting either the view of Plato, with his focus on the quest for the perfect form, or that of Aristotle, with his recognition of the importance of human desires in shaping decisions, but in integrating them. Based on that argument, our job as water resources planners and managers is to translate recent academic work on decision-making under uncertainty into a structured, analysis-supported, stakeholder-driven participatory process involving the full spectrum of stakeholders in a river basin and not simply interactions between the decision makers themselves. The process has been designed to facilitate negotiations among water managers and stakeholders holding distinct opinions as to the definition of a successful outcome—analogous to Plato's philosopher–kings and Aristotle's middle class, respectively—leading, it is hoped, to broad and stable agreements.

Whereas this quest clearly requires the participation of well-trained scientists and engineers, it helps if at least some have been exposed to the perspectives of the social sciences and humanities. So, when you get the urge to broaden your general education, why not consider taking a class in philosophy?

Floods Can Damage Careers in Communist Countries

Andreas H. Schumann

"Did you or did you not send floods to ruin our trout and cheese?"

The first part of my professional life I spent in a communist country, the German Democratic Republic (GDR), also known as East Germany. Despite its name, it never had a democratic government during its 40years of existence. In 1985, I was director for water management in one of the seven large water management authorities of this country. In this position, I was also responsible for the management of several reservoirs, located in the eastern part of the Harz Mountains. The local operators of these reservoirs had to open the bottom outlets once per month to test them.

Once, we ran into big trouble doing such a test. A trout-producing fish farm was located downstream of the reservoir. During the outlet test, water with a high content of suspended sediments spilled into the fish farm, killing all the fish. In the GDR, this was a clear case of economic sabotage, and somebody had to be responsible for it. The dreaded state security visited our authority. All persons involved in reservoir operation were interrogated. My colleagues and I were accused of flushing sediment from the reservoir, in the best case, and of harming the national economy in the worst case. The personal consequences of both accusations were not very exciting to think about. Preparing to confront the state security, my team and I were searching for the reasons of this exceptional event.

As mentioned, the bottom outlets were opened regularly, and no such effect was observed ever before. We had a closer look on discharge data at the inflow gauge of the reservoir. In the evening before our test there was a short but very intensive rain event in the reservoir catchment, resulting in a flash flood flowing into our reservoir. According to the land use conditions, we assumed that the inflow was loaded with sediment. Assuming full mixing of the inflow with the water already in the reservoir, it still could not explain the high load of suspended sediments at the outlet. The experts from the state security realized this and thus did not accept our arguments.

Then I remembered my studies in hydraulic engineering. If the load of suspended sediments was high, the density of the inflow could result in a short-circuit flow in the former riverbed within the reservoir. We contacted a specialist in hydraulics, working in a research institute, and he made some computations that showed the assumed short-circuit flow could happen in reality. He demonstrated that such a process could also result in hydraulic conditions within the deeper zone of the water body of the reservoir that could mobilize additional sediments. Obviously, this would explain the high rate of suspended solids at the bottom outlet the next day. Later, he used his computations for a case study in a popular German textbook of hydraulics, which was published in 1989. The state security and its experts accepted his clear scientific argument. We survived this crisis without negative consequences for anybody in our water management authority.

Floods were risky for my career in East Germany in several ways. In the summer of 1988, the socialist children's organization was organizing a large festival in the largest city of the East German district where I was the responsible water manager. In preparation for this event, a new building was proposed to be located on a floodplain within the city. I disagreed with this plan because of the high risk of flooding and argued against it. As the meeting was very important for propaganda purposes of the Communist Party, my objection became a political problem. The highest levels of functionaries of the district were involved in the planning. I was cited to appear before a commission, which consisted of officers of the state security and the army as well as from high-level representatives of the Communist Party. At that time, employees of water management authorities had to wear grey uniforms. I had a rank similar to a lieutenant colonel in the army. I decided to come in plainclothes to demonstrate that I was not supporting the government's position. Of course, in front of the committee was not the place to speak about flood risk, probabilities, hydraulics, or other good reasons to avoid a building at this site. However, after several years in administration, I had an idea of how to justify my decision. In this country, the highest steering committee of the Communist Party, the Politburo decided everything, including flood protection. Unfortunately, such decisions were not published. However, because of my job, I was aware of the existence of long-term plans for flood protection measures for this city. In the state planning system, such measures were part of the flood protection planning for the total river basin and for the whole country. I contacted some colleagues in other districts to get the information about any decisions of the Politburo about flood protection. I was lucky enough to find somebody who provided me with the date and file number of such a decision about the flood protection planning for the country. Now I was well prepared for the examination by the inquisition. It started with a monologue about the political relevance of my decision and all the problems originating from it, given by the district secretary of the Communist Party, who was chairing this committee. After that, I had a chance to revise my decision, which was obviously based on a bourgeois way of thinking. In my reply, I said I was very sorry to be forced to contradict the district secretary and then explained that the highest committee of the omniscient Communist Party decided to improve flood protection in this town, and I was even able to cite the date and file number of the resolution. Everybody was impressed by this information, as nobody was informed about such a resolution before. Of course, nobody was able to refute such a strong argument. After a few minutes of silence, the chairman said, "If there is a decision of the central committee, we cannot decide anything else." I was dismissed, but evil looks followed me. After the reunification of Germany, I was informed that the resolution I used in this power-play really existed, and the city I was protecting was even explicitly mentioned in it.

My last personal flood risk dates back to the summer of 1988. At this time in East Berlin, the capital of the GDR, there was a shortage of a special kind of expensive cheese. It became a topic in the council of ministers, and the minister for agriculture and nutrition blamed the minister for environment for this shortage. Following the arguments, it came out that I was the cause for this shortage! What happened? The cheese boxes, which were made of cardboard, were produced in a small factory located in my district. On a Friday evening, the cellar of this factory was flooded by a short, local flood event. There was no chance to replace the lost cheese boxes, the cheese could not be packed up, and the leading members of state and party missed their favorite cheese because of my failure to provide adequate flood protection.

In the state planning system, such flash flood events were not considered. The ministry sent a commissioner to prepare the indictment. A few weeks before, I had prepared my escape from the country, so I was relaxed. The factory was located in a part of the district where a former classmate of mine managed the regional water office. I was his boss, but we had (and still have) a good personal relationship. Now the commissioner from Berlin came to blame us for this flood event and the resulting losses of cheese boxes. The situation was rather amusing, and we had to grin several times. This annoyed the commissioner, and he announced hard, personal consequences against both of us. Knowing that such consequences would be nothing in comparison with the consequences of a failed escape, I did not worry about his threats. However, it was interesting how our impeachment was formulated. Every factory at risk from flooding was obliged to prepare a document describing how they would prevent damages in case of flooding. These documents assumed the existence of an early warning. Short flash floods were not considered. This is a general practice also today. In our case, the document for the cheese factory was unfortunately outdated by two years. My staff had not checked the updating, and I was responsible for this deficit. Because the damaging event differed significantly from the assumptions in the flood protection plan, there had to be a guilty person that caused the cheese shortage in Berlin. And that person was me. I have never seen the final report of this inspection, but I am sure it was not very complimentary to me. However, it did not matter. I left the country several months later, and this episode is now just one of several curiosities of my professional life.

How Water Probing Changed My Life Forever

Andreas H. Schumann

In 1988, I lived and worked in East Germany—the GDR. But five days after my 35th birthday, I spent my first night in a West German prison as a guest of the American secret service. It was the end of an eventful day that fundamentally changed my life and that of my family.

But let's start with the beginning. I grew up in a small industrial town in the eastern part of Germany. My parents were well-situated medical doctors, but nobody in our family had sympathy for the ruling communist system. As I was more interested in mathematics and physics, I studied those subjects together with hydrology and water management at the Technical University Dresden. After getting my Ph.D., I started work in a local water authority. I was responsible for the upkeep and operation of a 35-gauge hydrological observation network.

After three years, I became a deputy director for water management at one of the seven large regional water management authorities in the GDR. In this position, I was responsible for several departments and had close contacts to the scientific center of hydrology and water management of the GDR. I cooperated with internationally known East German scientists such as Alfred Becker, Dieter Lauterbach, and Stefan Kaden. At this time, Stefan Kaden was working at the International Institute for Applied System Analysis (IIASA) in Austria. His contract ended in 1986, and the Ministry for Water Management recommended me to succeed him at IIASA. I was deeply impressed by the working conditions at this institute and had a great hope to go there. But things changed rapidly.

One of my three brothers applied for permission to leave the GDR permanently with his family. I was prompted to cut off all contacts with my brother. I refused to do this. This decision ended my scientific and administrative career in East Germany. What to do in this situation? I had a family, was 33 years old, and had no chance to further my scientific career in the east. I discussed this situation with my wife, and we decided to make a sharp break from the communist state. To do this, we had to leave the country and go to the west. To apply for permission to leave the country legally would be dangerous and useless. In my former positions in water management, I was aware of the catastrophic environmental situation in the GDR, which was handled as a state secret at this time. Under these conditions, I had to develop my own plan to leave the state by crossing the Iron Curtain.

To do this, first I had to create an opportunity to defect. I stepped back as a deputy director and got another lower-paid position where I was responsible for water licensing, wastewater treatment, water quality, and hydrology in a district located in the southwest of East Germany at the state border to Western Germany.

There were several activities of the water management authority ongoing at the border, most of them dealing with the quality of small rivers crossing the border. I was informed that some members of my staff who were politically reliable got special permission to work close to the border. However, how could an unreliable person like me obtain such permission? Here, I was very lucky. There was a short list of people who had permission to work at the border. With my background, I could never get my name on this list. But in the summer of 1988, I saw this list on my desk. I just put my name on it! I was aware that this was illegal, and in the next weeks, I was very uncertain if the state security would discover this crime and arrest me. Maybe it was caused by the summertime or maybe I just had good luck, but nobody became suspicious.

The next task was to find a reason to make use of this permission to get close to the border. It had to be done quickly, as the list of permissions could be checked any day. Again, I was lucky. There was a request from West Germany to take special care of the water quality conditions of a small river flowing from east to west to protect some rare freshwater species that were living in it in western Germany. This river formed some hundred meters of the border between both parts of Germany. I explained to my director why I urgently needed more information about the water quality conditions of this river and requested permission to take some water probes there. It was a great surprise for my director that I got permission to do such work, but finally he sent me to the border. Nobody can imagine the feelings I had on this day, saying goodbye to my wife and to my two small daughters. I had three options for this day: if the attempt to escape failed, I could die or the state would send me to prison for several years; if I was successful, I would not see my family for many years and would have to start a new life as a homeless person in a new environment. The last option, which I excluded for myself, was to do nothing—to go to the border and to live in East Germany.

It was a cold and foggy morning when I came to the border. I had to cross several control lines and fences. Two noncommissioned officers of the National People's Army of the GDR joined me, both armed with their submachine guns. The entire time, I had to be very relaxed, speaking with them in the hope of developing a more personal relationship that might make it more difficult for them to kill me. We finally crossed the last barrier, and we were standing in front of the Iron Curtain, but were still in eastern territory. It was an open area, and in the middle the small river formed the border. Now was my chance to run, but at that moment I had a very bad feeling. There was nobody at the western side who could see me, and I felt that it was too dangerous to attempt an escape at that moment. What to do now? I had studied 50-year-old maps (new ones were not available) and had some ideas about the local situation. That is why I suggested we go a bit further upstream, where the river formed the borderline between Czechoslovakia and West Germany. It was forbidden to cross the border to Czechoslovakia, but I was able to convince my guards that it was necessary.

Fortunately, between both places, a small tributary flowed into the river, which gave me a good reason to go with my guard some hundred meters upstream into the territory of Czechoslovakia to take another probe. I stepped into the river, took my probe, and had a strong feeling that this would be exactly the right moment. I was very calm. I gave my water probes to my guards and started running through the river in the western direction. At the western bank, I jumped out of the water and started running through a small grove. My guards shouted at me and threatened to shoot. It was definitely the most dangerous moment in my life, and I was aware that my life could be close to its end. I heard one of the guards running after me and getting closer. I left the grove and ran in an open field. I had hoped that he would not follow me to avoid conflicts with West Germany, as we were several hundred meters in West Germany already. I was right; he did not follow me and did not use his submachine gun.

I ran to a small house, looked at the plate of the car in the yard to be sure that I was in the western part of Germany, and knocked at the door. The young family living there was very surprised to see me and to hear where I came from. I called my wife to say that I was alive and in the western part of Germany. The family phoned the police, and two police officers picked me up. Before I left, this family gave me all the money they had at home. The border police brought me into an office, gave me new clothes and something to eat, and examined me. What they were most confused about was the money the young family gave me. It was a considerable amount, and the police sent an officer to ask them if and why they gave it to me. The answer was, "Because this man needed it." After nearly 30 years, I am still in contact with this family.

At the end of this long day, the German border police brought me to the American secret service, which was located in an old villa. There, I spent the evening with an American GI (the tallest man I've ever seen) and a lady wearing dark glasses all evening. We were watching TV as the GI tried to explain to me the rules of American football and baseball (I never will understand them!). I also tasted peanut butter for the first time. All in all, it was a nice evening. However, when I wanted to go to bed, he brought me into a prison cell in the cellar of the building and closed the door behind me. The next morning, we had our breakfast together, they turned me over to the West German authorities, and I started my new life. In March 1989, I became chief engineer at the Institute of Hydrology and Water Management of the Ruhr-University Bochum in northwestern Germany, where I became a full professor 12 years later.

To avoid open questions: One year after I became a refugee the wall was broken down, in 1989, and at Christmas my family was with me in Bochum. I was often asked if I regretted the escape just one year before the Iron Curtain was falling. No, I never had any doubt that it was the right decision at the right moment. I still enjoy every day of my freedom and the fact that it was the result of a deliberate decision.

One Talk Leads to a Regional Program

Uri Shamir

"One minute I'm giving a talk, and the next I'm helping design Holland's water system."

In 1977, a mathematician at the Dutch National Institute for Drinking Water (RID, Rijksinstitut voor Drinkwater Voorzeining) read a 1968 paper I had written with my consulting engineer colleague, Chuck Howard, on "Water Distribution Systems Analysis." The paper described a novel modeling approach and computer program for simulating water distribution systems. The mathematician invited me to deliver a talk on that subject at RID and give them the computer program.

I traveled to Holland to talk about water distribution systems and to deliver the computer program. While we were dealing with this topic, the discussion expanded to more general issues regarding urban water systems. A broader group of experts joined us, and the discussion further expanded to regional water resources management. At some point, the group leader asked the deputy director of RID to join our discussion. It soon became clear that the experts were interested in the possibility of hiring me as a consultant on an ongoing project involving the development of an integrated drinking water system for the province of South Holland. The need for such an integrated system emerged from concerns about environmental degradation in the dune area along the North Sea shore. These sand dunes served two important functions—as a source of clean groundwater and as a prime nature and recreation area.

Over the next five years I served as consultant to the RID project, traveling back and forth from my home in Israel. I directed the systems analysis framework and optimization modeling for capacity building and operation of a regional water resources system for the province of South Holland. We had to deal with six competing objectives: cost, water quality, public health, reliability of supply, damage to nature, and energy use. A main driving force for the project was the damage to nature in the recreational dune area caused by water brought in from a river with a heavy load of nutrients and infiltrated through surface ponds to augment the groundwater. Because of the nutrient load, the natural vegetation was being replaced by an unwelcome vegetation of burning nettles, thereby reducing the natural and recreational value of the dune area.

Identifying the tradeoff between damage to nature and unit cost of supplied water was among the greatest challenges of the study: how to measure damage to nature due to the growth of burning nettles as caused by the ponding and infiltration of river water. The leading ecologist, a professor at the University of Leiden, was recruited to help us deal with this challenge. He, along with his students, not only collected ecological field data but also helped us convert these data into a "value function" that we used in our models.

This was the early 1980s, well before the advent of personal computers and their graphics, so presentation of results in the multidimensional objective space was a huge challenge. Representatives of the ministries and the provincial government supporting this study had to be convinced that the tradeoffs among objectives were realistic. This required us to not only quantify them in meaningful terms but also create visual displays of the tradeoffs among the competing objectives. That task is no big deal today, but it was then. The final report of our work was submitted in 1983.

The recommended plan included recharging the groundwater directly through wells, rather than using surface infiltration, thereby reducing the damage to nature. The plan also recommended increased use of groundwater in other parts of the province, and of the Rotterdam water treatment plant, and adding some infrastructure. It was approved and adopted by the government.

In December 2017, 34 years later, I received a surprise email from a few Dutch colleagues who had participated in the study. They wrote, "This afternoon we discussed what can be learned for the future from key developments in the past. The evolution of forecasting for policymakers and society came up. Your essential constructive inputs and particularly the simulations and optimization techniques you introduced into [what we call the "Zorgen voor Morgen" (Caring for Tomorrow) period (1985–2010)] were mentioned as breakthroughs. Although decades ago, your important contribution is still much appreciated."

This experience working with experts, stakeholders, and decisionmakers prompted me to promote the idea that the models we develop and use are important platforms for disciplined discourse among experts, stakeholders, and decisionmakers, not merely generators of computer outputs.

Trying to Keep Lake Balaton Blue

László Somlyódy and Pete Shanahan

"This is a sick lake."

The International Institute for Applied Systems Analysis (IIASA) in Laxenburg, Austria, is a think tank conceived in the early 1970s by US President Lyndon Johnson and the USSR Premier Alexei Kosygin to foster

East–West understanding. In 1978, a four-year collaborative research focused on Lake Balaton in Hungary began, involving IIASA, the Hungarian Academy of Sciences (HAS), and the Hungarian National Water Authority (NWA). The goal at the beginning was to develop ecological models for protecting the water quality of shallow lakes.

In mid-August of 1982, IIASA held its closing workshop on shallow lake eutrophication near Lake Balaton in Hungary. During previous weeks, the weather had been hot and dry, and as a result, the lake water was getting greener by the day. Journalists came one after another and raised unpleasant questions: Was there an accident? Is the water toxic? How long will the possible toxicity last? What can be done? Politicians at the workshop did not dare to talk. Most of them quickly disappeared and sent replacements. Panic. They offered no policy support or further research funding.

Was the workshop a failure? We learned somewhat later that the algal biomass concentration in that massive algae bloom had never been observed before. People were shocked; they were not informed before and had not realized that the lake was so sick.

Lake Balaton is a physically unique waterbody. Although large (about 75 km long by 8 km wide), it is extremely shallow, averaging only 3 m deep (the German fitting name is Platten See, plate lake). The lake is located in a limestone basin, and very fine limestone particles form its bottom. The shallow water is constantly stirred by the wind, keeping the fine limestone in suspension and causing dynamic changes in biology and many water quality components. It is a delightful place to swim, sail, and windsurf—most bathers are never in water over their head. It is a recreational and cultural resource treasured by Hungarians, with many summer homes, hotels, holiday camps, and wineries along its shores. The income from the tourist industry is about half of the country's total.

What did we know about Lake Balaton when we started our study that lead to the workshop? First, the lake's dominant algae species was completely unknown. It took months and months for the biologists to discover that the algae species was familiar in Brazilian and African waters. How the algae got to Europe via Australia is another story. And what will be its future paths? Second, the peculiar features of the algae in question were also unknown. As we learned, it is able to fix nitrogen (N) from the atmosphere and phosphorus (P) was available from the sediment in the bottom of the lake. We were surprised to see that this tricky species formed spores that survive the winter in the sediment and grow again the next year. Third, there were huge debates about which nutrient, P or N, or both, should be controlled. Views often changed, depending on research funding promises. (Yes, money seems to influence us.) Interestingly, the Ministry of Agriculture launched a program to monitor nonpoint nutrient sources. Their conclusion was—believe it or not—that diffuse pollution improves surface water quality. Fourth, there were limited experiences available on converting science results understandable to politicians in Hungary.

The region of the lake was associated with little infrastructure to control discharges of wastewater and nutrients. Massive fertilizer applications in agriculture and large-scale animal farming jointly contributed to a 10-fold increase of nutrient loads to the lake over half a century. One of the objectives of our research was to identify options that could effectively reduce the risks of algae blooms and improve the quality of the water in the lake. Possible control options included upgrading existing wastewater treatment plants, building new ones, sewage diversion, reduction of non-point sources, land use management, dredging, detergent control, and others. The preparation of the policy plan raised a number of questions. Do we know how to start? What should we control, N or P loads? What is the internal P load from lake sediment? Which form of loads should we reduce (dissolved, particulate, etc.; point and nonpoint sources; sewage and agricultural loads)? Do we have sufficient data and knowledge on nutrient balances and lake rehabilitation measures? How about control alternatives and costs? Who will react to what we find out, and how?

By the summer of 1981, professionals felt that conditions would lead to an "ecological disaster" in Lake Balaton if the summer was hot and wet. Hence, we focused our research strategy on how to avoid that potential catastrophe in that lake. The project became at once practical, specific, and fundamental.

By the end of 1981, we completed important analyses and offered answers to the questions addressed. Our work included the creation of a database, a lake ecological model, a nutrient load model, and a eutrophication management optimization model. This latter computed alternatives, cost implications, and scheduling actions (i.e., all those data needed for decision-making). We were ready to contribute. Was there an interest to accept this knowledge? We decided to organize a joint policy meeting at IIASA to present what we would recommend at the final workshop in Hungary the following year.

We invited six Hungarian vice ministers (it took us half a year to agree on all the details) in charge for various institutional aspects of the Lake Balaton problem. The meeting took place with simultaneous interpretation. There were open discussions and useful contributions. At some point during these discussions, the Hungarian interpreter gave me signals and sent me a slip of paper. He wanted to talk to me. He said, "I was listening to what the Hungarian ministers were whispering to each other: 'This is a useless meeting, too much talking about uncertainties, if this were the case, no decisions can be made.'" As the chair of this meeting, I ordered an unscheduled break. During that break, I changed a few slides and the tone of the presentation associated with uncertainties. Apparently, it worked. At the end of the meeting, the participants judged it a success. We got the attention of those who could do something to improve the quality of the lake, and we were prepared to show them how.

Taking into account what we learned in this 1981 IIASA workshop, we repeated our presentations at the 1982 workshop, attended by various ministers and journalists in Hungary. Given the lake conditions at that time, which were setting records in algal biomass, one would think our research results would be viewed as being useful for policy making. But as previously mentioned, this became an issue no minister wanted to tackle. Was it a failure of communication and persuasion on our part? It was certainly a surprise.

Watching the pitfalls in communication among various actors, we decided to rebuild our language completely. We tried to avoid the use of technical words—primary production, biomass, Chl-a, adsorption, limitation, P or N control, and so on—that led to endless debates and misunderstandings among scientists and decisionmakers. The institutional setting was rather complicated: six ministries or ministerial-level agencies representing urban development, agriculture, transportation, technology development, the Hungarian Academy of Sciences, and so on, three counties, three river basin authorities, many municipalities—and more than 50 others without well-defined tasks and responsibilities. Our communication challenge was to make things understandable to all these participants. In other words, we had to keep it simple.

We found that people still remembered what the color of the lake's water looked like 10 or 20 years ago. This offered us an opportunity to define future water quality goals in terms of the past years: for example, we would like to achieve a trophic state as it was at the early 1970s. We defined three states denoted by A, B and C: the actual state at that time, state A (early 1980s); the state associated with the early 1970s, state B; and the early 1960s, state C. Now, the objectives: state A, the water quality should remain unchanged (i.e., no further deterioration); state B, trophic state improves to what it was in the 1970s; and finally state C, further gain in water quality to reach what it was like in the early 1960s. This simple translation worked. It made a country—decisionmakers, scientists, lay personnel , and journalists—happy. They, remembering what used to be and possessing a knowledge of at least the first three letters of the alphabet, could understand and communicate with each other.

Actions were taken and monitoring was implemented that survived the turbulent political regime in 1990. By 1995, the total phosphorus load was reduced by about 50%, achieving state B. Today, we are at about 70% and close to state C. What have we learned? It is easy to destroy an ecosystem. Its recovery—assuming it can ever recover, given all the scientific and institutional uncertainties and hysteresis—takes much more time, money, and luck. Furthermore, we learned how to communicate with the decisionmakers: keep it simple and help them visualize what they want.

Part V

Adventures in the Americas

Water Crises Are for Everyone

Benedito (Ben) Braga

"There's nothing risky about "dead" storage!"

Brazil is a country with high diversity in many respects. It is the eighth largest economy in the world but has a large contingent of very poor and a small number of very rich people. The same is true with regard to water. The country has 12% of the world's freshwater resources, but 70% of this water is concentrated in the Amazon basin that hosts only 7% of Brazil's people. The northeastern region holds 30% of country's population but only 3% of its water. Southeastern Brazil, our focus in this story, is in the Paraná Basin. This basin includes the most industrialized and urbanized region of Brazil. Water crises in this basin adversely impact the local population and its, as well as the nation's, economy. Mitigating those impacts can challenge both the government and society. This story is about one of those crises.

The Location of the Crisis

The southeastern region of Brazil encompasses three of the most developed states: Sao Paulo, Minas Gerais, and Rio de Janeiro. The population of Sao Paulo State is 43 million and contributes 32% of Brazil's gross domestic product (GDP). The Metropolitan Region of Sao Paulo is located in the eastern part of the state. The natural water supplies in the region are insufficient to meet the demands of the metropolitan region's 21 million inhabitants. Back in the 1970s, a major interbasin transfer was implemented to bring water from the neighboring Piracicaba river basin through an 80 km complex scheme involving storage reservoirs, tunnels, canals, and a major pumping station. This is called the Cantareira system, and it provides almost 50% of the demand (73 m^3/s) of Sao Paulo.

A state water utility company provides water to this metropolitan region. The water supply system operated by the utility serves practically the entire population that lives in the formally urbanized area of Sao Paulo, where one can find meters in virtually all households. However, this is not the case in many of the informal settlements, where roughly 10% of the population lives. Most people living in the informal settlements have illegal and unpaid access to potable water through precarious and wasteful distribution systems formed by a bundle of small-diameter plastic tubes connected to the mains.

Several barriers prevent the water utility from entering these irregular settlements to provide regular services. Some relate to judicial disputes about the land ownership (e.g., people have invaded private property), others to environmental restrictions (e.g., people have invaded protected areas), and still others—the majority—to the impossibility of installing water supply and sewage collection systems in neighborhoods where there are no streets to bury the pipes. This unorganized urbanization process began in the 1950s, when only 30% of the Brazilian population lived in the cities, and has grown worse, especially in the early 1970s with the intensification of the industrialization process. In a few years, the urban population swelled by 70%, but the infrastructure growth was much slower, resulting in several problems. Among them is the large quantity of untreated sewage being dumped into the local water bodies. Thus, any new water sources have to be located far away from the demand center. Despite all these difficulties, practically all people living in the Metropolitan Region of Sao Paulo have access to potable water in their households, either through formal or informal connections.

Water and Politics

Beginning in 2014, water levels in the Cantareira system started to go down. Summer torrential rains that normally hit this region from December to February did not come as expected. By the end of 2014 the inflow to the Cantareira system was 50% of the all-time low discharge along 83 years of record. Associated with this totally unexpected hydrologic situation (the estimated probability of such an event is 0.004) in October 2014, we were having elections for governor of the state. The current governor was running for reelection and the opposition party was using the water crisis as part of their campaign, claiming lack of planning to cope with the problem. They were not successful and Governor Alckmin was reelected.

I was following this crisis very closely from my office at the University of Sao Paulo. One day, coming back from a meeting in Brasilia in December 2014, I received a call from the Secretary of Communication of the State of Sao Paulo, saying the governor would like to see me in his office. It was a Sunday afternoon and I was to fly that night to Cairo, as president of the World Water Council, for an international event hosted by the Ministry of Water of Egypt. I told him it would be impossible for me to attend the appointment, but he was very insistent. So, I went to the governor's office with my suitcase. I thought the governor would ask me some advice on how to face the challenge of the water crisis, but just at the beginning of our conversation he invited me to serve as Secretary of State for Sanitation and Water Resources.

He was very generous to give me freedom to invite professionals to serve in key positions at the offices of the Secretary of Communication, at the State Authority for Water Resources and at the State Water Utility. This was our dream team—well-known top academics and experienced professionals that were ready to work hard day and night. Without them, dealing with this water crisis would have been much more difficult.

Facing the Challenge

When our team took office in January 2015, the Cantareira system was operating below its minimum operational level. In January 2015, the Cantareira system storage was just at 5% of maximum storage, including the dead storage that cannot be released from reservoirs without pumping. The water levels were several meters below the intake elevations.

When the water level was approaching the minimum elevation necessary for the free flow through the intake of the upstream reservoir, the water utility installed a set of floating pumps and built channels and cofferdams in order to pump water up from the dead storage elevation to a provisional upper pond built adjacent to the intake. This simple engineering solution stirred a public debate fueled by the media about alleged health hazards for the people that would drink water from the "dead" storage. In fact, this was a false problem because a bottom gate had been for many years continuously releasing water from the reservoir to downstream to ensure the minimum environmental flow along the natural waterway. Nevertheless, it was necessary to implement a strong media action to explain that there was no risk to the health, including changing of the name "dead storage" to "technical reserve."

Besides using the dead storage, interbasin transfers (i.e., interconnections between sectors of the region) and associated hydraulic pressure relief valves were implemented. These interconnections allowed some 3.5 million people to be served from other supply systems, thus reducing the demand of water from the Cantareira. The pressure relief valves decreased the leakages, which amounted to 42% of the total water savings during the crisis. We understood that these infrastructural works were fundamental but not enough to cope with the challenge. While engineers were doing their work, the government of Sao Paulo and the utility launched a strong media campaign encouraging conscientious use of water. This involved numerous public meetings, lectures, and the distribution of posters and leaflets at condominiums, schools, commercial establishments, and homes, making people aware of the importance of saving water.

A nonstructural measure that resulted in significant savings was the granting of bonuses to those who reduced their water consumption and the charging of a contingency tariff to those who wasted water, all in an attempt to restrain demand. At the beginning of the crisis, a discount on the water bill was granted to those who saved water (i.e., to those who decreased water consumption, compared to their own average, by at least 20%). This program got support from the majority of the families. More than 80% of the consumers indeed reduced consumption. Of these, 49% received discounts in 2014 and 70% in 2015. The net effect was a lumped reduction in the potable water production, which corresponded to around 19% of the total water savings during the crisis. A penalty was imposed on those who did not understand the severity of the crisis and increased their consumption. Some 20% of consumers were in this situation.

These economic incentives persisted until the end of the crisis in 2016. The good news is that consumers changed their behavior. Even without the incentives, today, in 2018, the demand is 15% less than it was before the crisis. This past summer was very dry again, but thanks to the infrastructure and demand management implemented, no one felt at risk in the Metropolitan Region of Sao Paulo.

A Lock and Dam Story: Part 1. Background

Ben Dysart

In its February 1976 issue, *Readers Digest*—an influential general-interest magazine with a huge circulation—ran a feature article on the great controversy involving replacement of "Locks and Dam 26" (L&D 26) on the Mississippi River at Alton, Illinios, about 17 mi north of St. Louis. L&D 26, a major component of the nationally important Upper Mississippi River inland navigation system, was plagued with structural deficiencies. The US Army Corps of Engineers Civil Works Program proposed to replace two existing locks—one 600 ft and one 360 ft—with twin 1,200 ft locks at a 1970 estimated cost of $400 million.

Through their lobbyists and public relation folks, the inland navigation industry vigorously promoted the imminent failure of the existing L&D 26 that would block ship movement of grain down to the Port of New Orleans for international trade and ship movement of fuel oil up from New Orleans for power plants and industrial boilers in the Midwest. The barge/inland navigation industry claimed the failure of L&D 26 would produce unacceptable chaos, disruption, and catastrophic financial impact on the nation. This was part of the campaign—call it "air cover" for the US Army Corps of Engineers if you care to—to expedite design and start construction of the new, upgraded navigation infrastructure without delays from any congressional action. The Corps maintained that it had "implied authorization" for replacement; therefore, they argued, they need not seek and wait for any additional authority from Congress to spend $400 million.

Actually, it was a "stealth" inter-modal transport war between inland navigation and the railroads. The inland navigation industry wanted to greatly increase the capacity and throughput of L&D 26, which they saw as the choke-point on the Upper Mississippi for barge traffic. This increased capacity would cut costs and induce movement of substantial cargo, and revenue, from rail to water.

In plain English, "cutting costs" meant reducing unacceptably long (to the inland navigation industry) delays in transiting the locks. The railroad industry countered that it was all greed on the part of their competing mode—the barges—and that the $400 million cost to replace L&D 26 was simply the first piece of an unholy, multi-billion-dollar navigation conspiracy with the Corps of Engineers, and their mutual friends in the Congress, to replace *all* the locks and dams in the Upper Mississippi over several years, as the choke-point would simply shift to another of the 25 or so Mississippi River dams and their locks upstream from L&D 26.

Thus, L&D 26 was the battleground. Billions of taxpayers' dollars were at stake to substantially increase the competitiveness and profitability of one mode of transport at the expense of the other. One mode was a customer of the US Army Corps of Engineers; the other wasn't.

The mighty bull elephants—and their surrogates—were in mortal combat in the jungle! And each was supporting its proxies as well as all its lobbyists and public relations folks. Both rail and barge sectors were funneling a lot of money through front organizations to support them and undermine their opponents. This was a familiar scenario for high-stakes federal water resources development projects and systems over many decades in the United States.

Meanwhile, back in a Washington courtroom, it took a federal district judge exactly a month to rule that fear for the system-wide impact of lock expansion was a reasonable concern. The pretense that plans for bigger locks were unconnected to plans for bigger channels was, the judge admonished, "unworthy of belief." On September 6, 1974, the judge suspended the design work on the replacement for L&D 26 at Alton. First the Corps would have to meet the requirements of the National Environmental Policy Act (NEPA). If Congress wanted an oversized double lock, the Corps would have to figure the cost—to the railroads, to the taxpayers, to the environment—of bigger locks and dams up and down the system.

Quickly, the Corps went back to work on a supplemental environmental impact statement (EIS) that considered the wider impact of bigger locks but arrived at the same two points: first, the danger to the environment was minimal; second, the old timber and concrete project was a hazard that should be replaced. Critics became apoplectic. *Reader's Digest* called big locks "a multibillion-dollar rip-off." A *60 Minutes* television report implied that "big business" had hoodwinked "big government" into building a Taj Mahal.

One would assume that with such unbelievably high stakes for Midwestern cities and all of agriculture in the Upper Mississippi basin—and for the Corps's reputation for technical and project management excellence—there would be an impeccable, compelling, state-of-the-art analysis of the alternatives and their benefits and costs to produce a sound benefit/cost ratio. However, recall that a judge had found the Corps's initial efforts at expedited project justification for their $400 million replacement to be "unworthy of belief." Would their second try involving sophisticated systems analysis techniques—including mathematical modeling and state-of-practice computer simulation—be more rigorous and credible?

On May 6, 1976, Senator Gaylord Nelson (D-Wisc.) requested that the text of the *Reader's Digest* article, "Big Dam Decision at Alton" and an exchange of letters between the Corps and the managing editor of the *Digest* be printed in the *Congressional Record* of the Senate.

The first paragraph of the February 17, 1976, critique of "Big Dam Decision at Alton" from Maj. Gen. J. W. "Jack" Morris, deputy chief of engineers, to the chairman and editor-in-chief of *Reader's Digest* expressed concern over what he considered as misstatements and oversimplifications of a very complex issue to the point of leading readers away from the fundamental problem that existed at Alton, Illinois.

In his reply to General Morris on March 15, 1976, the editor-in-chief stated that without seeing any indication of any misstatements, oversimplifications, and inaccuracies, he considered Morris's response as evidence of the fact that the *Digest* took an extremely complex subject and gave its readers a lucid, accurate, and responsible account of the issues involved.

Gen. Morris got his clock cleaned by a serious journalist and executive. I knew the head of the Public Affairs Office (PAO) in the chief's office who handled public relations, press releases, speech drafting, and so on, and I predicted that Gen. Morris would lose out if he depended on one of their—the PAO shop's—letters.

Enter the water resources systems analyst. In the spring of 1975, as a young associate professor of environmental engineering at Clemson and coordinator of Clemson's Water Resources Engineering graduate program, I was recruited to a fascinating billet in the Office of the Secretary of the Army (OSA) in Washington, DC.

I reported to the Pentagon in September 1975, just as the massive intermodal battle over L&D 26 reached a fever pitch. My job title was Science Advisor to the Assistant Secretary of the Army for Civil Works (ASA-CW). The Corps's Civil Works program was long and widely viewed as essentially "the construction arm of the Congress," pretty much beholden to the powerful chairmen of the Public Works and Rivers and Harbors authorization and appropriations committees and subcommittees. Their members were mostly from the Deep South, particularly the Lower Mississippi River Valley. In addition to managing the design and construction of major water resources development projects, the Corps targeted projects that generated windfall benefits to individuals who had acquired and maintained what's euphemistically referred to as "productive access to the political process," as indicated by the keen interest of powerful members of both houses of Congress and their staffers. Most of these members believed in federal programs to promote the well-being of their constituents. After all, that is how they keep their congressional jobs, and eventually their leadership positions.

The assistant secretary, Hon. Victor V. Veysey, was the "President's man" overseeing the US Army Corps of Engineers Civil Works Program. He was the first in that new position, and the Corps was not at all happy to have to report to a political appointee. He was viewed by the Corps as one who insisted that they be a part of the executive branch instead of their comfortable role of reporting essentially directly to their sole "customer" (i.e., the powerful members of Congress and Senators who gave them their projects and the dollars needed to implement them). His oversight role didn't please these important members of Congress either. It was sort of like the process of getting a wild horse accustomed to the bit and bridle, and perhaps similarly traumatic! But to me, he was a good guy and a fellow civil engineer who valued and would use credible technical and policy-level advice. Working for him was a very satisfying experience for me.

A Lock and Dam Story: Part 2. My Involvement

Ben Dysart

I left my Clemson University job and reported to the Pentagon in September 1975 as Science Advisor to the Assistant Secretary of the Army for Civil Works (ASA-CW) just as the massive intermodal battle over the Mississippi River Lock and Dam 26 (L&D 26) replacement project reached a fever pitch. My first assignment upon arriving at Office of the Secretary of the Army (OSA) was to liaise with a working group of the Young Presidents' Organization (YPO). This organization consists of presidents and CEOs of companies with a substantial revenue who are under 40 years of age—a very high-capacity population. Members of YPO volunteer their considerable top-management expertise to assist their counterparts in government with complex projects and issues, such as L&D 26. I was to be "the secretary's man" working with a YPO workgroup.

A couple of months before I reported to work, I received a large volume of L&D 26 documents and was requested to digest them thoroughly prior to arrival. I went through all the piles of documentation, the design memorandum, the reports, the appendices and boilerplate, the endless pages of computer printouts, the history of L&D 26, the correspondence, the testimony, everything, over and over.

As I recall, the YPO working group studied the complex and opaque L&D 26 situation for a while but never got close to understanding what was going on. The working group wrote a report to the assistant secretary, were thanked, and moved on. This lack of understanding was not by accident. A project approval process runs more smoothly if the planning and economic justification of the project intended to deliver targeted financial benefits to one's constituents (i.e., constituents of key members of Congress and of the Corps or any federal agency) is "complex and opaque." Intentionally obscuring the political/economic/moral trade-offs, the operative objective function, weighting factors, and actual constraints that are operative (and those that are not)—and introducing material that fogs up the process—actually tends to help facilitate the project approval and funding process in Washington, DC.

But this assistant secretary *still* wanted me to help him figure it out.

I knew this meant carefully picking my way through the

- "hard, but necessary, official public-sector and private-sector political/economic/moral trade-off processes"—my term;
- long-standing process of "externalizing" one's (or one's supporters') costs onto society or, conversely, creating targeted public transfer payments; and
- "Iron Triangle" consisting of the Corps, Congress, and the congressional supporters in the states and districts from where they come.

I knew the political drivers were not (1) the "official" calculated total benefits to the public and (2) the "official" calculated total costs to the public. Rather, they were the dollar benefits and costs to whatever entity or individual(s) had caused the project to successfully wind its way through the congressional authorization and appropriations process. The dollars actually driving the real process may or may not be a part of the "official" project benefit and cost figures. Such figures were not infrequently chicken entrails and tea leaves (i.e., smoke and mirrors).

Two analyses of project alternatives and their public benefits and costs were needed: (1) one for open presentation to the congressional authorizing and appropriations committees and subcommittees to justify the expenditure of a lot of public money and (2) one that yielded the politically "correct" answer.

One of the traditional beauties of water resources projects is their effectiveness in targeting benefits to those with productive access to the political process. This phenomenon is not limited to the water resources project business but is a traditional characteristic of *all* federal programs that are part of the numerous "Iron Triangles" in the federal government. But the US Corps of Engineers / powerful congressional interests / congressional supporters "Iron Triangle" has long been an awesome machine to watch as it proceeds with its work.

The basis for Republican President Nixon's "Southern Strategy" that enabled him to have a majority governing coalition and prosecute the conflict in Vietnam was water projects in the South and the Lower Mississippi Valley. Water projects were the political "coin of the realm," a carte blanche in Vietnam.

Eventually, after reviewing everything I could find and interviewing a host of involved individuals, I requested that the St. Louis Corps District economist responsible for evaluation of L&D 26's benefits bring all his files to Washington and explain how they came up with their benefit number. This seemed to be pivotal. I told him to plan to stay for a week.

He gave me the horror story about how the sandy foundation had been eroded over time from under a portion of the structure and there were now "dangling piles" waving to and fro. I probably laughed (we *both* did). Then he told me the saga of the failing, spalling-off concrete on the bottom of the dam. I asked for evidence, so the district sent a diver down to confirm it; the diver cut an 18 in. section of old exposed rebar where the concrete had spalled off and hand delivered it to me. A great paperweight and a horror story indeed. The objective of the campaign for L&D 26 improvements? The project must be constructed as requested and funds appropriated without delay to avert a disaster for the nation and its people! Such urgent messages warrant judicious use of the snicker and smell tests.

I certainly wasn't interested in the structure's failing or in shutting down the Upper Mississippi, but my gut told me something was amiss. Several days and a couple of Excedrin-worthy headaches later, the district economist had doggedly stuck to the official Corps talking points; but somehow it still didn't hang together for me.

On Friday, I thanked him, and suggested that we go out for a few beers that afternoon before he headed to the airport. And we did so.

After a few beers, I opined, "And further, I don't think all those hundreds of pages of Monte Carlo simulation printout and the nonsense conclusions in your oh-so-thick economic-justification appendix had anything to do with generating or confirming your project benefit number." He just looked at me with no expression.

I continued: "I consult for a world-class dam builder, and I teach a water-resources systems analysis graduate class, and one technique I cover is Monte Carlo simulation. If a student team submitted junk like that to me as a semester project, I wouldn't accept it." This seemed harsh, but I had to say it.

He shrugged, paused, and casually said: "All this proves is that your grad students at Clemson are better than those who did this for us."

Bingo! So, the hundreds of pages of fancy systems analysis were, being excessively charitable, simply irrelevant boilerplate. Fake, sham, smoke and mirrors, and BS were used to *imply* that state-of-practice systems analysis had produced a credible, unimpeachable project benefit number. This is a big deal, because historically, benefit figures have not infrequently been "jiggered" or back-calculated. I've had district and OCE planners and economists share such stories off the record. Benefit and cost analyses are sent back to be redone until they're "right."

"Here's what I think," I said. "You simply multiplied three numbers: (1) the projected number of tows transiting the locks in a year, (2) the estimated average hourly cost of a tow waiting to transit the locks, and (3) the estimated reduction in wait time for a tow with the 'replacement' locks. A simple back-of-the-envelope economic-benefit analysis. Right?" He just shrugged and smiled, then said, "Didn't figure anyone in DC would notice or figure it out." And it nearly worked.

"Or *want* to figure it out? Or *care*?" I added. He shrugged again. He had a plane to catch, and I sat there, satisfied but exhausted: A "back-of-the-envelope" economic-benefit analysis for a $400 million taxpayer-funded project, and they didn't figure anyone would notice, figure it out, or care? Was this state-of-practice water-resources systems analysis circa the mid-1970s in the big leagues? It makes one wonder.

A Lock and Dam Story: Part 3. Lessons on How Government Works

Ben Dysart

"You want it bad, you get it bad."

How does one make sense of this Mississippi River Lock and Dam 26 replacement experience? It would be difficult for an independent, credible, well-informed, nonfinancially conflicted, wise person to do so merely by looking at the L&D 26 cards spread face up on the table. But this will help a lot: I will never forget what was conveyed to me sotto voce in 1975 by a distinguished Pentagon economist colleague; his four underlying truths/"decoder-ring" rules for understanding serious official government decision making in (1) the water resources project planning, design, construction, operations, and regulatory arena, (2) protecting human health, (3) the use of the nation's natural resources, including those on public lands, and (4) environmental protection. For understanding projects and decisions that, on their face, seem totally irrational, counterintuitive, illogical to "the reasonably prudent man," and seemingly in neither the nation's nor the public's interest:

- "That is no accident." There are reasons behind every political decision.

- “What must be done will be done” and “What must *not* be done, will *not* be done.” Pay attention to what the political process wants and doesn’t want.
- “You give a little to get a lot.” Otherwise you may not get anything.
- “You want it bad, you get it bad.” The political process gets what it wants.

All this was true then for L&D 26 in the mid-1970s, true now, and likely true forever regardless of the controlling political party. These four rules explain a lot. Now you have the key to unscrambling the (1) disingenuous, dissembling politician, (2) bureaucratic regulatory agency, and (3) congressional double-speak. The same rules apply at the state and local levels. You can see through the political decision-making glass more clearly. What was crooked may now look straighter, or vice versa. Of course, “crooked” and “straight” depend on whether one agrees with the objective function crafted to be optimized and the constraints imposed and those excluded. The validity of these four laws has been repeatedly demonstrated throughout my career not only in the public sector but also in Corporate America, in full-time consulting, in leadership of NGOs, and in the academy.

Was my peek behind the veil vis-à-vis L&D 26 economic benefits and systems analysis a fluke? A unique situation? I can’t say definitively, but in another very controversial several-hundred-million-dollar Corps dam project under review at the same time, I requested a senior planner from the Planning Division of the Office of the Chief of Engineers (OCE) to come to my office to discuss and explain its justification. I told him I wasn’t totally opposed to dam projects and acknowledged that the last good, solidly justified dam might not yet have been conceived, designed, and constructed in the United States. But, I said, given this and speaking as a civil engineer and a water resources engineer, this project the Corps was pushing so aggressively for a few members of congress and their financial contributors sure seemed like a sorry embarrassment. Surely the Corps could come up with a better water development project for those several-hundred-million public dollars.

The senior OCE planner thought a moment, nodded, and said, "Well, sir, I can see why you would view the project as a bad, inefficient public water resources development project with seriously jiggered total public benefits to exceed public costs—a loser. But look at it another way: It's a great, highly targeted, effective winner as a private land-value enhancement project for those who have acquired 'seriously productive access to the political process.'" I thanked him and said that now the pieces fit nicely.

Incidentally, this senior planner returned from lunch one day to find the chair of the House Appropriations Committee, Mr. Whitten himself, sitting at his (the OCE planner's) desk—an absolutely unheard-of thing. The chair said he understood that a certain project in which he was terribly interested had been modified. The planner struggled to act composed and explained that, yes, several design modifications had been made that substantially enhanced public benefits and reduced costs as well. Mr. Whitten was not amused. He stood, put a wagging finger in the planner's face and explained firmly that the project was exactly as he wanted it when he got it authorized on Capitol Hill. He made it clear that if he ever wanted anything changed, he'd let the Corps know how he wanted it changed. Then the pillar of fire walked out, leaving a singed planner who now better understood his job.

Back to the other controversial dam project: A bit later—after the OCE planner explained that a shaky taxpayer-funded water resources project might be a politically attractive transfer payment in the form of enhancing the value of private land—I accompanied the assistant secretary to the middle of nowhere for a luncheon of typical local chamber of commerce economic boosters. The big, green US Army helicopter was met by the half dozen men who would realize some millions of dollars each from the several-hundred-million public dollars Corps dam project. Private land-value enhancement was not a listed purpose of the project, and those to receive the millions in windfall benefits would not cost-share.

Inefficient? Yes and no, depending on which objective function, which constraint set, which set of variables, which coefficients, and so on you choose to examine.

There's all sorts of water resources systems analysis being done, from (1) solid and credible, and I've done some, to (2) pure "advocacy science"—fake, dodgy, or "sponsor-directed" science. The four cardinal rules can help you understand what you're looking at:

- Rule 1: "That is no accident."
- Rule 2: "What must be done will be done."
- Rule 3: "You give a little to get a lot."
- Rule 4: "You want it bad, you get it bad."

To these chiseled-in-granite laws, I might add a couple of "Ben-isms" from my course in water resources planning:

- Rule 5: "Life is a series of trade-offs."
- Rule 6: "Life is played out in the gray zone."

These last two overlap a bit.

Here's the take-away: State-of-the-art or -practice water resources systems analysis tools such as mathematical modeling, optimization, and simulation are essential to properly inform public-policy decisions vis-à-vis complex water resources development projects *when* the intent is an efficient and effective outcome legitimately serving the long-term, broad public interest.

This is how it's supposed to be within the academy: by the book.

However, nonpurely textbook dimensions that *dominate* in the real world are ignored or minimized at one's own peril. Dimensions of any project include the social, political, institutional, economic, moral, and ethical.

However, cutting through the good intentions, happy talk, and BS, the *realistic* practitioner of water resources systems analysis must appreciate that there are intents *other* than "an efficient and effective outcome legitimately serving the long-term, broad public interest" oftentimes desired by "those who count" (i.e., those who own the decision-making process or those who control those who own it).

The realistic practitioner must understand that oftentimes there are several 900 lb gorillas in the room that preclude “an efficient and effective outcome legitimately serving the long-term, broad public interest” including (1) hard, but necessary, official public-sector and private-sector political/economic/moral trade-off processes, (2) the long-standing process of “externalizing” one’s (or one’s supporters’) costs onto society or, conversely, creating targeted public-dollars transfer payments, and (3) the “Iron Triangle.”

The realistic practitioner must understand the four cardinal rules listed earlier that can help one see what’s going on.

“Those who count” (whoever they may be) typically know the characteristics of the outcome they want; and this can, in *some* instances, be to serve the public interest. More commonly a somewhat, or greatly, different outcome is expected.

All this greatly affects the mathematical model or system conceptualization to be optimized or simulated. It also affects what is to be optimized, the constraints, and other factors considered or ignored. And it’s conceivable that there could be a couple of rather different models and programs: one for public consumption and one for internal use serving the goal of those who count.

Lessons I Learned from Western Movies and TV Shows

David T. Ford

Water resources system engineering (WRSE) is a branch of civil engineering in which solutions to water resources planning and management (WRPM) problems are found by systematically applying quantitative analysis methods. Those methods account for interconnected and integrated hydrologic and hydraulic, infrastructure, ecologic, and human processes, thus aiding formulation, evaluation, and selection of solutions. Because of the breadth of WRSE, those of us who work in this area can honestly claim that we have learned from many sources lessons that have been helpful in our work.

An important source of lessons for me has been Western movies and TV shows. Westerns center on the life of an American cowboy in the latter half of the nineteenth century. The movies and shows commonly depict commission of a crime, followed by pursuit, capture, and punishment of the criminal. The punishment often involves a gun fight. Well-known stars of these movies and TV shows include John Wayne, Clint Eastwood, Roy Rogers, Gene Autry, and others I name following. Following are five helpful lessons I learned from Westerns.

Welcome Others to Your Campfire

In a common scene in a Western, cowboys are sitting around a campfire at the end of a long day on the trail. A stranger approaches in the dark and weapons are produced quickly. Commonly, after some initial skepticism, the stranger is invited into the circle and becomes part of the group. (He may even pull out his guitar to serenade the other cowboys.)

I've learned from this the value of getting beyond skepticism to collaborate with professionals outside our normal circle of civil engineering colleagues. For example, for a study of long-term operation of Glen Canyon Dam, I was able to join a geographer, anthropologist, zoologist, sociologist, limnologist, and biologist "around the campfire" to consider options and impacts beyond what civil engineers would have seen, leading to a better operation strategy.

Don't Criticize the Cooking If You've Never Been in the Chuck Wagon

In many Westerns, a group of cowboys herds cattle from a Texas ranch to a market, moving across the US plains. Such a cattle drive takes days or weeks, so the group is accompanied by a cook and a mobile kitchen—a chuck wagon. A cowboy who complains about the food from the chuck wagon often is forced to trade places with the cook, sometimes after harsh words and a fist fight. The newly appointed cook quickly comes to appreciate the unseen difficulties of the task, learning that the theory of good cooking on the trail differs from the realities of practice. The lesson I learned from such scenes is this: I shouldn't criticize solutions to a WRPM problem unless I fully appreciate the problem, unless I am prepared to implement my solution in the real world, and unless I am willing to take the criticism aimed at "the cook."

For example, it's tempting for me to review California's newly published Central Valley Flood Protection Plan and imagine how I could use a particular optimization method to improve the plan. However, unless that approach will represent the critical hydrologic and hydraulic, infrastructure, ecologic, and (especially) the complex human processes that affect flood management decisions in California, the elegant "meal" I imagine preparing if I were in charge of the chuck wagon may fail in practice to satisfy those who consume the results. Details matter; details are difficult to understand without firsthand exposure and experience, and details are often difficult to represent properly with an analytical tool.

Never Forget the Water

A common plot element in Westerns includes a grueling chase through a dry desert, with one cowboy evading another. Predictably, one of the cowboys fails to pay attention to his need for water to drink, then runs short, worsening his situation.

Those Western stories remind me to pay attention to water, particularly the principles of hydraulics, water chemistry, aquatic biology, and elements of water science, as I seek solutions to WRPM problems. Sure, it's convenient for me to use a simplified routing model to answer reservoir operation questions. But important flow conditions are not always represented well with a simplified model. A WRPM plan I identify with sophisticated WRSE methods, built on a foundation of an oversimplified representation of the watershed, channels, ecosystem, or water management features, is unlikely to be optimal in reality, and it may be infeasible.

Identify the Good Guys

In many Westerns, it's easy for the viewer to distinguish between the good guys (they wear white hats) and the bad guys (black hats). But in some Westerns, as with some WRSE problems, it is not easy to identify that which is good. For example, in the Western movie *The Man Who Shot Liberty Valance,* the good guys are not wearing white hats. That creates a dilemma for the character played by Jimmy Stewart. Only when he realizes how to measure goodness is his path forward—although difficult—clear to him. This movie taught me to take care in how I measure what's good as I evaluate alternative solutions, since appropriate evaluation often is the most difficult task in WRPM.

The Flood Control Act of 1936 permitted the federal government to participate in projects "if the benefits to whomsoever they may accrue are in excess of the estimated costs." Accordingly, I and others developed WRSE models with objective functions that measure the net economic benefit of alternative solutions, and we used those models to find "the good guys." The Water Resources Development Act (WRDA) of 2007 led us to broader goals in WRPM, including protecting and restoring functions of ecosystems, mitigating unavoidable damage, encouraging sustainable economic development, avoiding unwise use of floodplains, reducing public safety risk, seeking solutions that fairly treat and meaningfully involve all people, and taking a watershed approach. Now, to be successful, I must do as Jimmy Stewart did, gain an understanding of which objectives are important, find ways to measure contributions of alternative solutions to those objectives, and analyze trade-offs among conflicting objectives to find a compromise.

Finish What You Start

We systems engineers are deliberators. We like to deliberate alternative solutions and alternative paths to developing alternative solutions. But in reality, progress will not wait on us. I learned from the 1952 movie *High Noon* that I must stop deliberating at some point, make a difficult choice with incomplete data, offer an opinion, and finish what I start. In *High Noon*, the marshal (Gary Cooper) must make a difficult choice between alternatives: leaving town with his fiancée (Grace Kelly) or staying to protect the citizens from bad guys. (In this case, the bad guys are recognized easily.) Cooper's character agonizes, debates, deliberates, and finally makes a decision. We viewers are left to decide if he made the optimal choice, but he finished, and he did so in time to make a difference. Decisions we make about WRPM are similar. We often have imperfect data, unclear objectives, and model results about which we are uncertain. I learned from *High Noon* that I must act with these to meet a real schedule that demands action. If my action improves the condition and does so in a timely manner, I may be *successful enough*.

Bonus Lesson

A bonus lesson I learned from Westerns is to keep a good sense of humor about my work in WRPM. Certainly, our work as civil engineers is important, and it demands serious attention, somber thought, and careful analysis, design, and operation. But even in the darkest episodes of the TV series *Gunsmoke,* Chester Goode (Dennis Weaver), Marshal Dillon's friend, kept a good sense of humor and added levity. I've learned that friendship and a good sense of humor will contribute much to our success solving WRSE problems.

This story has been adapted from my article in the *Journal of Water Resources Planning and Management* Vol. 144, Issue 5 (May 2018) ©2018 American Society of Civil Engineers.

Inventing GIS

Walter M. Grayman and Richard M. Males

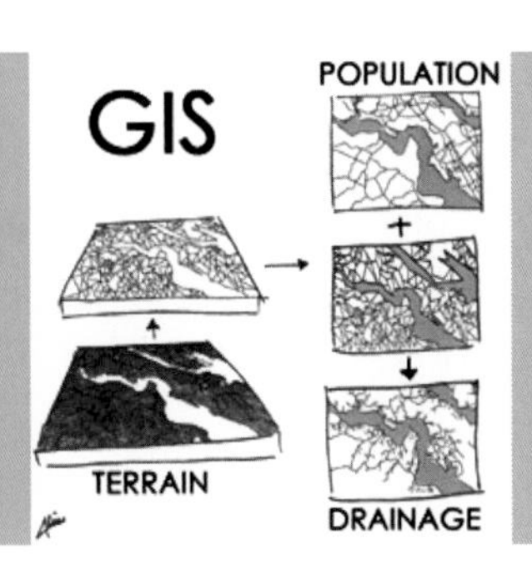

Although this may sound a little like Al Gore's claim that he "invented" the Internet 45 years ago, a small group of us actually did invent GIS (geographic information system), or at least the idea of the use of GIS as an engineering tool. Here is how it happened.

In 1973, a small group of us at Engineering-Science headed by Dr. William E. Gates were working on a regional water quality management study of the lower James River in Virginia. As we developed the conceptual approach for the study, we knew that we wanted to be able to evaluate a large number of alternative regional wastewater management plans under a range of growth and population scenarios. Ultimately it required doing planning-level designs for regional sewer and treatment systems under alternative growth scenarios, estimating the cost of each alternative, and examining the impacts of the resulting effluents on the receiving stream network. This was no small task given the size of the region (6,000 mi^2), the number of scenarios that we wanted to examine (hundreds), and the time schedule (6 months). If you were given this task today, certainly the idea of the integration of engineering models with a geographic information system would be the obvious solution. However, in 1973 the term GIS was just an obscure reference to a Canadian mapping system, CGIS, the Canadian Geographic Information System.

To get the ball rolling on how to address this somewhat intractable problem, our group of several engineers, a geologist, and a surveyor got together in an all-day brainstorming session. We all agreed that the solution would involve the development of a spatial database representing terrain, land use, demographic data, and physical data such as soils, as well as the development and application of sewer design and costing models directly integrated with the database. We also knew that elevations and slopes were key factors in designing sewers and for other engineering tasks such as calculating runoff. At that point in time, existing mapping systems used regular rectangular grids to store elevation data. This did not seem like a good solution for us, because this resulted in a series of flat plates rather than a continuous terrain representation that we needed for our engineering work. In addition, the constant resolution (all rectangular grid cells of the same size) would either fail to capture detailed terrain or result in a large number of small grid cells that was computationally challenging with the computers of the time.

Several members of the team had surveying experience and recognized that a network of triangles was potentially a better solution since a plane is defined by three points. We proceeded on our planning by building a faceted triangular network to represent terrain, generating and representing the stream drainage network as the triangle sides, and then storing other spatial characteristics in the triangular network. At the same time that we were developing the concept of triangular networks, two other researchers around the world were simultaneously and independently developing similar networks. In GIS jargon, the use of a triangular network for representing terrain became known as a TIN (Triangulated Irregular Network) structure, a concept that is frequently used today for representing terrain.

The development of GIS and its application to the lower James River basin is documented in a paper published in the ASCE *Journal of Hydraulics*. This is likely the first paper in an ASCE journal on use of GIS and the earliest reference to GIS in the Web of Science.

Following the completion of the Virginia project, Engineering-Science had no interest in pursuing this area, and as a result, most of our group of engineers and planners left and formed our own company and continued further developing and marketing the tool. The tool was formally named ADAPT for Areal Design and Planning Tool. Over the next 10 years, the tool was refined and widely marketed. In most situations we found ourselves in competition with a small land-use planning firm called ESRI. Though ESRI won most of the contracts we competed for, ADAPT certainly enjoyed a modicum of success. The State of Ohio selected ADAPT as the basis for a statewide data base system that they called PEMSO (Planning and Engineering Management System for Ohio).

The State of Kentucky adopted ADAPT to represent several river basins; it was applied in studies in parts of several other states around the country, and it was used as an interface for the Hydrologic Engineering Center for use with its HEC-1 and HEC-2 models.

By the mid-1980s, ADAPT had not created enough traction to be commercially successful, and the company folded. However, many of the concepts live on today. GIS has become big business, and ESRI, our chief competitor then, is now a $4 billion business. The TIN structure is widely used today, and many other researchers have independently adopted the concept of embedding a stream network in the TIN structure for hydrologic analysis. Importantly, the idea of "data-driven modeling," championed by the ADAPT team as a key organizing concept, in which an underlying spatial database is used to both supply data to a variety of complex models and to capture output information for analysis and display, has become a standard and useful approach. We look back with satisfaction on having invented it, and from almost a totally clean slate. Most of the basic concepts we introduced were new. That is extremely satisfying in this crowded research world where most research involves small, incremental steps to what has been done before. In short, doing this was fun! Knowing that it was, and continues to be, used makes it even more satisfying.

Rescuing a Beautiful River

Neil Grigg

"We'll get to the bottom of this algae bloom, Rufus!"

Up until the mid-1970s, my career in water resources was mostly technical, working on various university projects and in consulting. Then, I had an opportunity to work on a project that opened my eyes to another world. It was one of those fortuitous opportunities that only comes along every once in a while, and it introduced me to work and opportunities that were more expansive and consequential than any other I'd had.

How I got to this opportunity takes a little explaining as it involved a series of steps. While at Auburn in the mid-1960s and then later at Colorado State University I worked as a student and faculty member on projects funded by the newly formed Office of Water Resources Research (OWRR). That's the program that spawned the state water resources research institutes (WRRIs), most of which are still in operation. The interdisciplinary OWRR projects and the state and local government connections they provided seemed neat, so I got more involved over time. Eventually this led to my being appointed as director of the North Carolina WRRI in 1977. This required me to move from the west to the east, into a very different water world.

After arriving in Raleigh, it was not long before I heard about research on the Chowan River in northeastern North Carolina. Mostly, I heard about it from my predecessor, David Howells, who had been a terrific director of the NC WRRI and who deserves credit for transforming it into one of America's best WRRIs. A major issue at that point was algae blooms in the river and how university research might help determine its causes and remedies.

Our work on this issue led to my being asked to take leave and assume a job in the NC state government as Assistant Secretary for Natural Resources (ASNR) within the Department of Natural Resources and Community Development. NC had a tradition of rotating academics through state government, at least in natural resources. Compared to moving purely political people in, this was a breath of fresh air, to say the least. The department has changed its name now, but the organization at the time included a suite of natural resources agencies. The Division of Environmental Management was the lead agency to handle water quality.

First-term Governor Jim Hunt was under pressure to solve the Chowan River problem because a couple of industries were being accused of polluting it in ways to create nuisance algae blooms. The governor decided to host a large stakeholder meeting to gather some facts and opinions, and I was directed to explain what was involved and what needed to be done. I had to scramble because at that time I knew next to nothing about it, but fortunately I could find people who did, so we got a presentation together. Right away I found out three things. First, the technical part was a long way from settled. People had monitored and modeled the river, but we didn't know what was causing the algae blooms. Prime suspects were a paper mill upstream in Virginia and a fertilizer plant in North Carolina. However, there was a lot of crop and animal agriculture in the basin, and other nonpoint sources could have contributed to this problem as well.

The second thing I learned was that people were really mad about it and wanted it cleared up fast. This was why the governor got involved personally, because it was a big issue for a lot of people in eastern NC, a resource-based region and part of his political base. He was from a small town in eastern NC and had been raised on a tobacco farm.

The third thing I learned was that we were between a rock and hard place because, without the evidence about what was causing the problem, you couldn't take effective action. Without such action, you couldn't satisfy the people who were upset. Gathering evidence takes a long time and involves money for scientists to do their research, and people are not willing to stand still for very much of that. So, we really had to do something.

The challenge to restore the river energized me and others in the department. After my presentation, I recall the governor turning to my boss, the cabinet secretary, and saying something like, "Howard, if we can't clean up a river, maybe we should hang it up." We were then sent back to the office to develop a plan. This involved a series of meetings to go over what we knew and to develop options. Although it was clear that the algae blooms were due to excess nutrients, whether it was primarily due to nitrogen or phosphorous and to which sources had not been determined. No one had identified a smoking gun to nail down which nutrient source was the primary culprit. Having a smoking gun would have been politically desirable, of course.

I found out quickly that I was walking on eggshells in trying to galvanize the bureaucracy to help in this. It's a long story for which the title should be *How Many Ways Can Old Bureaucrats Fool and Ignore a Short-term Academic in State Government who Lacks Much Authority?* You get the idea. Fortunately, the agency had younger technical people who really wanted to help, and we were able to prepare a plan, which mostly relied on provisions of the Clean Water Act.

This was in 1979, and the Clean Water Act was passed in 1972, so most provisions were still being implemented. The department had authority under various sections of the act to conduct studies, such as the 208 program that enabled us to inventory nonpoint sources. In 1979, the Chowan became the first river basin in North Carolina to receive the "nutrient sensitive waters" classification. The state's Division of Water Quality recommended reductions in the amounts of phosphorus and nitrogen that wastewater treatment systems were allowed to discharge into the river.

It was clear from this inventory that a fertilizer plant was a primary nutrient source. Some inspections had been carried out, and there were rumors about old fertilizer waste being buried on the site and leaching slowly into the river. Monitoring data were relatively inconclusive, but we were converging on the conclusion that the fertilizer plant was an important factor. At the same time, a papermill upstream in Virginia held papermill waste in holding ponds and released the waste rapidly so that it moved down the river as a big black slug, which drove the fish ahead of it in fear and infuriated local fisherman.

As a result of suspicions about the papermill and other activities in Virginia, we started a program of cooperation with the Virginia State Water Control Board to see if something could be worked out jointly for a total river basin approach. Because Virginia was the upstream state, it did not see immediately what it should sacrifice for the benefit of North Carolina, so we had to look for win-win cooperative strategies. One of the things we noticed was that the water supply for the growing cities in the Tidewater area was inadequate, and Virginia might need our help in facilitating project permits. Another issue was that a large amount of groundwater was being pumped to lower the groundwater table in the whole region, and perhaps we could influence that practice for the benefit of everyone. During this investigation, we learned that there was a large interbasin transfer out of the Chowan River basin to Norfolk that had been implemented as an emergency measure during World War II to help the navy, but it had never been subjected to review by North Carolina and was obviously affecting the river. Taken together, these were issues that could be lumped into a package for cooperation.

One of the main lessons I learned in all of this is about cooperation with local people. The town of Edenton, North Carolina, was a central player from its location at the mouth of the Chowan River. Right away I found allies in Edenton, mainly in the city government and the local news media. These became valuable resources to us as we sought integrated solutions to the problem. Nongovernmental groups were important too. In particular, a local leader, Al Howard, provided a lot of leadership in motivating and galvanizing local services to help us out.

Another important aspect of the program was seeking the US Environmental Protection Agency's (EPA's) help to bring a modeling team from the Chesapeake program to look at what we were doing to see if it made any sense. The governor helped us arrange this by making a personal request to the EPA regional administrator. The modeling team was competent, but they produced more questions than answers.

This experience taught me many lessons, and it certainly influenced what I taught my students upon my return to academic life. Models and data are helpful and, in fact, often critical in addressing the "what if" questions during debates over just what decisions to make. But absolutely critical are social skills and the ability to work with and lead people to any successful solution of any complex and contentious water management problem.

How a Civil Engineer Learned to Work with People

Stacy M. Langsdale

"That looks great!" "That looks unacceptable."

When I started my first job as a civil engineer, I dove into the work at my large drawing table, kept my head down, and didn't chat much, because I wanted to be a good employee—a good, hardworking engineer. I designed ditches, culverts, and grading plans to be built by people I would never meet, often on sites I'd never see. The work was interesting but thinking that would be my next 40 years of work depressed me.

I escaped to graduate school in search of new opportunities and adventures. I knew I wanted to do something in water resources, but what? Through fortunate circumstances and choice, my advisor worked in water resources policy and introduced me to an interesting question, "How can we communicate science in a way that decision-makers can understand and use the information?" I still use this to describe my career 20 years later.

For my master's project at the University of Nevada, Reno, I evaluated options to restore Walker Lake and the threatened Lahontan cutthroat trout that were struggling to survive in it. Walker Lake, a terminal lake (i.e., having no outlet) in western Nevada, was largely depleted and brackish (i.e., a salty puddle) after a century of river diversions, primarily for irrigation, to support a large agricultural economy in a dry landscape. Purchasing water rights for the lake was only an option if current water rights holders were willing to sell. As that seemed unlikely, we explored "wild" alternatives like desalination and cloud seeding as well. I developed a model, and the results contributed to an Environmental Impact Statement (EIS) for the Bureau of Land Management (BLM). But I also got to attend my first public meetings.

The modeling work was intriguing because many results were counter-intuitive. The conditions of having too many water rights on this terminal lake actually made typical solutions counterproductive. I discovered that water-saving approaches such as lining ditches would actually *reduce* the amount of water reaching the lake, because water that leaks into the ground can slowly flow to the lake, while water in the ditches can be diverted for human use. Desalination technologies require a lot of water, so that would really shrink the lake, making it even more vulnerable to drought.

I spent hours running simulations by myself, and I developed a solid understanding of the dynamics of the system. Although I shared the results both in the EIS and my thesis, I worried that these quantitative tables and written descriptions didn't convey all of the important findings I'd learned about the behavior of the system through observing model simulations.

- Another formative experience for me was meeting stakeholders at three very different public meetings: In the agricultural community of Yerington, upstream along Walker River, 150 angry farmers asked why we should save the ugly lake and reminded us that they put food on our tables.
- In a smoky restaurant with slot machines in the town of Hawthorne, built at the south side of the lake and now eight miles from the receded shoreline, community officials explained how they depended on visitors who fish in the lake and help Hawthorne's economy when they buy gas and provisions. For them, saving the lake and the fish was vital.

- In Carson City, state government representatives were professionally—not personally—invested, so the placid discussions focused on understanding the technical analysis and clarifying the process and schedule for the EIS (with one exception: a seven-foot-tall cowboy in a Stetson as wide as the doorframe stopped by to tell us we were all wasting our time because plans had already been designed in the 1960s to save *all* the lakes in the western United States—by piping water from the Columbia River!).

The experiences at these meetings only reinforced my concern that our communication of results was limited. Furthermore, I had read in the literature that you needed to involve the decision-makers in the development of the model if you wanted them to use it. I worried, "Would decision-makers understand our results and make the right decision?" As an engineer, I could calculate the "optimal" solution. But I was beginning to see that the only really robust solution (purchasing a *large* number of water rights) was neither politically acceptable nor economically viable. Would BLM take desperate measures and invest in alternatives with high risks?

Nearing graduation, I felt motivated to try again so I could involve stakeholders from the beginning. So, I proceeded to a doctorate program. This time graduate school was much different as I entered with a clear vision of what I wanted to do.

At the University of British Columbia (UBC), fortitude combined with (more!) good fortune got me into an interdisciplinary program, with a receptive advisor, and a situation that was ripe for a collaborative modeling approach: exploring the future of water resources in the Okanagan River basin.

Here, I stepped outside of my science/engineering box into a new world of social and behavioral studies. Engineering and science courses were straightforward—imparting tools and techniques for calculating the optimal solution. But planning and environmental resource management discussed how to approach problems in the real world—amid people's values and interests; navigating history, relationships, and personalities;

as well as needing to balance environmental, political, and socioeconomic considerations. In these courses we got to debate ideas and share our opinions!

I took a class on decision analysis, where a friend and I were the only engineers among planners. When we learned a tool that *quantified* people's preferences and *calculated* the group's preferred alternative, my engineering friend and I thought this was fantastic, but all the planning students cringed. It was a good lesson on the need to match the tools with the audience. It also reinforced that models and analysis tools *support* decisions, while *people make* decisions. People consider many factors when making decisions, only a few of which are taught in engineering school. Decision support tools thus need to incorporate what is important to those who will make or are affected by a decision. And this can only be achieved by involving stakeholders with a variety of perspectives and interests in the process.

So the Okanagan watershed became my testing ground for involving stakeholders in developing a decision support tool to explore their water future. With the urban population exploding and climate change increasing the need for more water for crops, the area was poised for an agricultural versus urban showdown. The research team had generated many separate pieces of the story, but would the region actually run out of water? If so, when? And what actions should be taken to prevent a water crisis?

Over the course of a year, I combined all of the team's research findings into one dynamic model while involving stakeholders over the course of five meetings. I had no training in facilitating meetings, so I just stumbled through the workshop design and execution by instinct. (I recall at the first meeting attempting to motivate them to participate by declaring: "I want to make sure this is a useful process for all of you, not just an academic exercise. If it's not useful, we can all go home." I think I did catch their attention, but in an unsettling rather than inspiring way.) I was fortunate to have help from others who were a couple of steps ahead of me in stumbling along this path.

When I jumped into this evolving field, researchers had a range of approaches to involve participants in model building. Have them watch you code? Build the model live during the workshop? Not realistic. Instead, we gathered local knowledge from them about the history of water in the basin, as well as their concerns. Later, we had them critique a simple model. At the last meeting, attendees ran the model themselves. Despite designing the interface as user-friendly as we could, it still took considerable time for participants to learn to navigate the (now complex) model to get it to test what they wanted. That ended up being inefficient and left little time to discuss the effectiveness of solutions. Regardless, the local planning board referenced the project in their strategic plan, so we declared this a success in translating science for decision-makers.

Engineering gave me skills to calculate technically feasible solutions. However, decisions are made in messy real-world contexts, so we must also consider political and social acceptability, economic viability, and the values of those affected. The more we can incorporate these factors and metrics into our modeling tools, the more useful they will be to decision makers. And we can only learn these aspects by engaging decision makers and stakeholders in the process.

So, I learned that a "good" engineer doesn't have to bury his or her head in solitary work. On the contrary, "good" engineers need to talk to decision-makers and affected stakeholders to ensure our work is relevant and useful for supporting decisions.

Managing Our Nation's Water Resources When It Rains a Lot: An Example

Deborah H. Lee

"Flood Management"

As the chief of the Water Management Division for the US Army Corps of Engineers (USACE), Great Lakes and Ohio River Division (LRD), I managed the largest system of multipurpose reservoirs east of the Mississippi. This was a challenging and daunting responsibility and especially important in the field of water management and flood control, where an exercise of poor judgment can literally mean a difference between life and death, as well as billions of dollars in damage.

By way of illustration, in 2011, I was called on to execute lower Ohio and Mississippi River flood control for what would become the largest flood on record. This responsibility entailed not only operating the USACE reservoirs, but by law, directing the Tennessee Valley Authority's reservoirs as well. It also required close coordination with the Mississippi Valley Division (MVD) and its operation of the Mississippi River and Tributaries system for flood damage reduction all the way to New Orleans, Louisiana, and with the Northwest Division and its operation of the Missouri River system.

As if this were not a sufficient challenge, many of the Great Lakes and Ohio River Division dams were in the top ten national dams in need of repair. Ranked first on this list was Lake Cumberland's Wolf Creek Dam—a dam unparalleled in its water storage capacity but also most at risk from a dam safety perspective. This dam stood at the headwaters of the Cumberland River system in a chain of reservoirs that are key to protecting the Mississippi River valley. The dam had been under repair since 2007, but by 2011 this repair was not yet complete.

As the great flood evolved, beginning in late February and lasting through July, I developed the technical flood fight strategy that included unprecedented coordination with the National Weather Service to issue record flood forecasts ten days in advance of the crest, and I led technical water management communications on a daily basis with the USACE Divisions, their districts, and the National Weather Service River Forecast Centers, exercising flood control operations on a continental scale. At the height of the flood, I worked 21 days straight, up to 14 hours daily, not only testing my technical expertise, but also testing my physical and mental stamina.

As the flood crests from the Missouri and Ohio Rivers raced toward their junction with the lower Mississippi River, it became apparent to me that without using the storage in Lake Cumberland with Wolf Creek Dam under repair, I would not be able to reduce the colliding flood crests sufficiently enough to prevent overtopping of the Mississippi River and Tributaries system levees, even with the planned operation of the Birds Point–New

Madrid Floodway. I brought in the Dam Safety Program to assess whether repairs had progressed sufficiently to an acceptable risk level. A miscalculation at this point would have had catastrophic effects. The division commander accepted the risk, and the end result was a successful operation, now memorialized in the book *Divine Providence.* Major General Peabody, 36th president of the Mississippi River Commission, wrote in my copy, "Please accept this . . . in grateful acknowledgment of your superb water management expertise, professional dedication, and outstanding judgment. The nation is in your debt."

We—my whole team—were lucky. What we did worked. This example highlights how important continued maintenance and repairs of our water infrastructure are to make sure our facilities are ready for potentially catastrophic events or circumstances, and how impactful our decisions can be to life and property. But no matter how well we engineers plan, design, operate, and maintain our infrastructure, the outcome of this, or any, flooding event could depend on something we have yet to learn how to control: how much it rains.

Into the Realm of the Patagonian Andes

Leif Lillehammer

In early 2009 I traveled back to Latin America, this time to work on a hydropower project in Aysen, which is part of Patagonian Chile. This is quite a different place compared to the tropics, where I seem to end up for most of my international consulting projects. Aysen is about 2,000 km south of Santiago and has a fantastic scenery and landscape. Huge glaciers reside up in the mountainous areas. I was to undertake a study on the impact of a series of hydropower reservoirs on the aquatic ecology and ecohydrology in the Pascua and Baker Rivers. Both of these rivers are heavily glacier fed, and glacial lake outburst floods (GLOFs) occur quite frequently. At least one occurred while I worked on this project. During the 2009 trip, we flew in a helicopter along both of these rivers, and also up to some of the glaciers that feed them. It was one of the most magnificent experiences during my career, even though being frightened by a strong gust of wind when hovering in that helicopter above one of the glaciers.

The areas around the Baker and Pascua Rivers are highly untouched and pristine, with low settlement density. Many NGOs and environmental professionals value this part of Chile with its river systems, opposing developments such as this proposed hydropower reservoir cascade. This is probably one of the reasons why plans for the development were abandoned later on.

Dealing with aquatic ecology again and looking into mitigation of peak operation of the cascades in both rivers, I revisited the work of Mark Bain, who had studied this all the way back in the 1980s but also published a comprehensive review on hydropower operation and environmental conservation in 2007. The Cornell influence seems to be with me wherever I go.

Managing Lake Water Levels

Pete Loucks

"That muskrat makes a compelling point."

Where I live in the middle of New York state in the United States, there is a large deep lake that is used mainly for water supply, recreation, and wildlife habitat. People that live along the lakeshore have located their docks and boat houses based on their expectation of lake levels during the summer and fall seasons. A dam at the downstream end of the lake controls the lake level and downstream flows, at least to some extent. The dam is operated by the New York Department of Transportation because it is a part of the New York State Barge Canal System that crosses the state.

One of the first studies I undertook in my water management career was aimed at deriving an operating policy for managing the lake's water levels, given the uncertain inflows that occur throughout the year and the need to prevent flooding both upstream and downstream of the lake. To get an idea of what water levels different shoreline owners would like to have, we rented a boat and traveled around the shore to meet them. Our goal, besides having fun, was to obtain from each property owner their desired lake levels over the year. We thought and expected there might be conflicts among different types of water users that included recreational boaters, swimmers, those living in the local flood plain that can be flooded if the lake level gets too high, and those who value the extensive wetlands for bird and other wildlife habitats. Indeed, conflicts among different types of users did exist, but what surprised us the most were those conflicts just among marina operators that service recreational boats.

The marina operators who seemed to be relatively prosperous desired high lake levels so that their customers could easily enter their docks to obtain gas, food, and other services and products. Those that seemed less prosperous preferred low lake levels so that they would have more work, and thus more income, repairing the damaged hulls of boats that had ventured into too-shallow waters. That was my first exposure to what we call multiobjective problems. Not everyone agrees on what is best.

Some 40 years later, I was engaged once again in looking at lake—and river—water level management. This time it was for Lake Ontario, the lowest of the Great Lakes, and the St. Lawrence River that flows from Lake Ontario to the Atlantic Ocean. This system forms part of the boundary between Canada and the United States. Its management, along with all waters flowing between the two countries, is the responsibility of the International Joint Commission (IJC). The IJC is a group of six individuals appointed by the two governments. I was part of a study IJC created to define an operating policy for managing water levels in that lake and river that would be more responsive to the health of the wetlands in the system than was the existing policy—a policy in use for almost half a century.

In addition to keeping muskrats happier (short for ecosystem health), we also had to derive a policy acceptable to the hydropower producers, recreational boaters, water supply providers, commercial navigation interests, and shoreline owners. Each substantially contributes to the economy of the region, and hence each is politically influential.

It was clear to us from the beginning of this study that for any new operating policy to be acceptable to the IJC, and thus to the governments of Canada and the US, all stakeholders on both sides of the border would have to find the new policy acceptable. This meant they had to understand it. They would have to trust those who were developing policies and evaluating the impacts of such policies. To us, this meant we had to engage these stakeholders in this study. I have never been in any study where more attention, time, and money were spent attempting to keep every individual and institutional stakeholder, both English and French speaking, in the entire lower Great Lakes and St. Lawrence River basin, informed. In turn, we had to ensure that we were fully and continuously informed of, and responsive to, their concerns.

Nevertheless, by the time our five-year, $20 million study ended, we failed to find any policy acceptable to everyone. Some stakeholders were simply not willing to compromise, and they let that be known to their political representatives in their countries. Some stakeholders trusted those spreading misinformation (that served their personal interests) more than what we were telling them based on our analyses. All this is not so surprising, but it did keep us motivated in trying to communicate our findings more effectively and convincingly in an attempt to gain the trust of all stakeholders.

A year or so after our study ended, the US co-chair of the IJC told me that because we analysts couldn't give him the monetary value of a muskrat, in either Canadian or US dollars, the IJC couldn't carry out a cost-benefit comparison that included ecosystem benefits along with those of the other interests. Thus, he said, the IJC commissioners could not make a decision based on our study. The end result: a large pile of documents and supporting data pertaining to each stakeholder group's interests, three recommended candidate policies as were requested out of the hundreds created and evaluated, many news media articles and records of stakeholder meetings and workshops (all presented in two languages), and a very interesting and educational but somewhat disappointing experience for us. We were appointed by the IJC to find a policy that worked for everyone, and we couldn't.

Some nine years later, one of our recommended, but slightly revised, operating policies was accepted by the IJC and by the governments of Canada and the US. Two years after that, it rained a lot, and the entire lake and river shoreline experienced substantial flooding. No policy would have prevented this flooding. But as expected, many shoreline owners blamed the new policy for causing the flooding and urged the IJC to reject the new policy. And as often happens, affected state and provincial governments have offered to pay at least part of those property owners' losses resulting from the risks the owners took by locating themselves close to the shores of the lake and river. These owners are certain to remain there and experience more flooding in the future no matter what operating policy is being implemented. And no doubt, when that happens, our taxes will help them recover.

Restoring the Everglades

Pete Loucks

The Everglades

Just when people think they know a little about how water systems behave, they should spend some time working with those responsible for managing the water flowing through the Everglades. The Everglades, located in the southern half of the state of Florida in the US, has been called a 60 mi wide river of grass. I've had the privilege of working there at various times throughout much of my career. I recall that when I first went there, water managers told me I should forget about what I learned from textbooks: "It doesn't apply here." True enough. I learned that in that flat area, water can flow in any direction, even in canals or rivers. The direction of flow depends on where it rains. Water can flow through this flat wetland at infinitesimally low velocities and still create a unique parallel ridge and slough (valley) topography. The interaction and exchange between surface and groundwater flow is continuous. Its ecosystem is unique. Even a tiny amount of phosphorus in the flow of water through this wetland can cause a change in the plant community of this ecosystem. This subtropical ecosystem is indeed unique in the United States and thus attracts many tourists to the region.

Drainage projects over the past decades have caused a steady degradation of this ecosystem and its extent. If this degradation is to end, water has to be managed differently from how it has been in the past, when the goal was to drain the swamp to enhance economic development. The South Florida Water Management District is responsible for managing water in this region of South Florida. This multidisciplinary group of very talented scientists and engineers have been my teachers.

One of my involvements with this district focused on defining how water might be managed so as to improve the health of the plant and animal species making up this ecosystem. Instead of trying to better understand and predict the behavior of individual animal or plant species, we chose to focus on the condition of their habitats. With substantial help from a variety of ecologists who had expertise in various components of this ecosystem, we were able to define the relative suitability of particular habitats as a function of water management decisions, such as the durations and timing of water flows and depths. We created a set of these functions related to the stage of development of various animal and plant species. I knew we had to include the habitats of the animals that people come to the Everglades to see, such as alligators, birds, fish, and panthers. But I learned it was also important for us to include in our analyses the habitats of species that no one comes to see, such as three different species of algae. Who but ecologists would have guessed! We also created habitat suitability functions for topographic features, such as tree islands, that the ecologists considered important. What we learned from the use of these functions in our analyses of various water management policies is that it is impossible to keep all species of this ecosystem in bliss—that is, in perfect health—all the time. Ecosystem science in the Everglades, in my simple view, is all about who is eating whom and when. What we were looking for was a water management policy that would keep the entire ecosystem resilient—when some are in stress, others are not, and vice versa. Water has to be managed to preserve all species and features of this ecosystem.

Our habitat suitability models were relatively simple, and thus relatively transparent to the governing board of the district. During one presentation of our results to them, the board chairman stated: "Finally we have something we can use." This conclusion apparently threatened those scientists who were getting considerable financial support for their efforts to model the behavior of various animals themselves, not just their habitats. These scientists were not happy, and they publicly criticized our work. Their objections diminished over time as they continued to get the funding they wanted to continue their more demanding and challenging research. That experience reminded me that we scientists are also in an ecosystem; when some are enjoying success, others may not be, but all need to be kept resilient!

Mistaken Objectives

Pete Loucks

"The goal is not to be cost-effective, it is to keep our jobs safe."

When most of us think of the state of New Jersey, we think of people, lots of them, in a generally urban setting. After all, New Jersey has the highest population density of all the states in the US. But in part of the state, the scenery looks more like a part of Vermont. This is the Raritan River basin, where New Jersey's largest reservoir by volume is located. The Round Valley Reservoir is a pump-storage reservoir that has a capacity at the spillway level of 55 billion gallons (about 210 million m^3) of water. To augment water storage, water is pumped into the Round Valley Reservoir from the South Branch of the Raritan River, making it available for later release and therefore increasing the so-called safe yield of the Raritan Basin System. Approximately 1.5 million people in central New Jersey rely on that system for their water supplies.

This Round Valley off-stream pumped storage reservoir is filled by ten 40 MGD pumps pushing water uphill through a 3.2 mi pipeline from the river to the lake. For example, in 2017, 7 billion gallons of water was pumped to the reservoir over about six months, raising the water level by about 10 ft. Almost all of the water that goes into the Round Valley Reservoir has to be pumped there. A set of reservoir operating rules govern actual reservoir releases, pumpages, and water-supply withdrawals.

Some time ago, I learned about this system and its operation. At that time, the operating policy was to keep the reservoir full of water. As soon as the level dropped below full capacity, the pumps were turned on to refill the reservoir. Then, of course, it would rain, and because the reservoir volume was at capacity, the runoff to the reservoir from the rain just flowed over the spillway.

Being young systems analysts, my smart students and I had the idea that maybe we could save the agency in charge of operating this reservoir some money while still maintaining acceptable levels of supply reliability. We thought we could do this by considering reservoir volume pumping targets less than the full reservoir capacity. If possible, this would permit the capture and use of at least some of the rainfall runoff, and thereby save pumping costs, which were substantial. So, with the operating agency's cooperation (and no doubt amusement), we tested the system response (both costs and reliability) associated with various storage targets over the historical river flow record. We indeed found a target level that would meet the then-current demands for water through the 1960s drought of record while saving about a million dollars a month. (We were hoping they might give us a portion of that savings!)

When it was time to present our results, we were invited into a room containing a long, fancy mahogany table surrounded by lots of soft chairs filled with agency department heads. After making our pitch, it seemed obvious to me they really didn't seem very interested in what we had to say. I asked them if that was the case, and their answer was yes. I assumed the reason they might not be interested in implementing our results was because they were not eager to do anything that might justify getting a lower allocation of money to operate the reservoir. When asked, their response was no—my assumption was wrong. Rather, they reminded me that they were in the business of supplying a reliable source of water to their customers, not saving money. If a drought indeed occurred that exceeded the severity of the drought of record in the 1960s, if they failed to meet the demand for water during that more severe drought, and if it was learned that they were reducing safe yield reliability just to save money, they would be heavily criticized. No doubt the jobs of some who were sitting around that table would be lost. The risk was not worth the savings.

Another experience I had at the beginning of my research career was to identify the influence of various block-rate water pricing policies on cost-effective capacity expansion schedules and strategies for urban water systems. I was working on this problem with a Ph.D. student from Winnipeg, Canada, so we used that city as a case study. Our approach was to not be constrained by our assumed demand function for water in the future, however uncertain such functions are. Rather, we allowed existing capacities to fail, and when they did fail, we paid the added cost of providing water from alternative sources. Thus. we included in our analyses the probability of having to pay for any measures needed to make up for such a deficit. This seems more like an approach an economist might consider as opposed to that of an engineer who typically assumes the demands must be met.

Soon after that study was completed and after my student received his Ph.D., I happened to be at a conference sponsored by ASCE. ASCE's president at the time was Sam Baxter, the then-current water commissioner of the city of Philadelphia. By chance, I ended up sitting next to him at a banquet. To make conversation, I told him of this study we just completed. He listened politely, and then when I was finished, he asked, "Do you know what would happen to me if someone in my city turned on the water faucet and nothing came out?" No doubt, the mayor of that city might be finding a new water commissioner.

The moral of these stories: Try to determine the real objectives (and constraints) before trying to satisfy them.

The Challenges and Rewards of a Career Managing Water

Lindell Ormsbee

I first became interested in water resources while living in the Panama Canal Zone as a child in the 1960s. As a military kid, I would frequently pedal my bike to the canal to watch the ships pass through the locks. I was mesmerized by how engineers working in the early part of the twentieth century could accomplish such a feat. What fascinated me the most was the fact that the canal was supported by lots of components that all operated together to make it work. These components included several dams, lakes, channels, pipes and pumps, roads and rail, water and wastewater treatment facilities, and even complete cities. In addition to gaining an appreciation for the science and engineering of water resources, I discovered that such systems must be designed and operated using a systems approach, that is, one that takes into consideration the interrelationships of the various subcomponents and the impacts from the perspectives of economics, political science, sociology, and public health. This initial real-life exposure to civil engineering and the concept of systems analysis continues to motivate me today as I pursue a career in water resources.

As an undergraduate civil engineering student at the University of Kentucky, I became intrigued about the nature of the flow of water in open channels such as sewer systems and in closed conduits such as water distribution systems. While working in the hydraulics lab, I discovered that sometimes water did not behave in ways that I expected. That motivated me to find out why. I learned what caused laminar and turbulent flow, sub- and supercritical flow, hydraulic jumps, and even something called a "water hammer."

While pursuing a master's degree at Virginia Tech, I wrote a computer program that could be used to predict the performance of hydraulic systems. Shortly after graduating from Virginia Tech, I got a job in a consulting firm putting those skills to work in the real world. One of the most rewarding experiences of my time in consulting was interacting with clients (mostly municipalities) and learning firsthand the nature of their water-related problems. What I quickly discovered was that many of their problems looked nothing like what I had seen in homework problems in class. They were much more complicated, and in most cases the information needed to address the problem was not provided. I had to identify and find that needed information myself! While certainly a challenge, it was always extremely rewarding to be able to solve a real-world problem and then see my solution implemented. Today, I can visit various communities in Kentucky and look up and see a water tank that I designed, visit a small dam that I helped renovate, or look down and see the manholes and catch basins of a storm sewer that I helped develop. I know that such solutions have truly improved people's lives.

After several years as a consulting engineer, I realized how much I didn't know, and that motivated me to pursue a Ph.D. at Purdue University. I wanted to better understand how things worked. Once again, my focus was on water resources and systems analysis. After graduating, I returned to the University of Kentucky, but this time as a member of the faculty. While at UK, I was fortunate to work with one of the pioneers of water distribution modeling, Dr. Don Wood. During that time, we developed a computer program called KYPIPE that predicted the flows and pressures in water distribution systems. Today, upgrades of that software continue to be used by engineers around the world. As a result of that partnership, I have had the opportunity to travel around the United States to work on water systems of various cities, including Phoenix, Kansas City, Detroit, and Atlanta.

One of the most fascinating projects I worked on was with Dr. Tom Walski, then of the US Army Corps of Engineers. This project involved developing a way to reduce the amount of energy used by the water utilities of Washington, DC, and several communities in northern Virginia, including the military base of Fort Myer and the Pentagon. Water utilities use a lot of energy, mainly for pumping purposes. We performed a study to see what would happen if part of the water system failed and how we could build redundancies into the system to minimize the impact of such failure. As part of that study, we conducted flow tests of hydrants right on the Pentagon grounds as well as at the old National Airport in Washington, DC. We got a lot of strange looks during our field work, but it was fun.

Another fascinating project I worked on was at the Paducah Gaseous Diffusion Plant in Paducah, Kentucky. This uranium-enriching facility was built in the 1950s during the Cold War. The facility is surrounded by security fences and armed guards as well as a large wildlife management area that blocks the view of the facility from the surrounding area. During its continual operation from 1952 to 2014, the facility used an enormous amount of energy (equivalent to the energy used by the city of St. Louis) and water (primarily for cooling) as part of the production process. The plant has a control room that looks like something out of a *Star Trek* movie. During the time the facility operated, significant soil and groundwater contamination occurred. Part of our job was to figure out where the contamination was coming from, how much there was, where it was going, and how to clean it up. We were also tasked to work with the local community to solicit their ideas about future alternative uses of the facility. This project was challenging because of (1) the nature of the site (which currently contains the largest concentration of spent uranium in the world), (2) the suspicion and concerns of the local citizens about the radioactive contamination at the site, and (3) the potential health impacts from the contamination, especially for children. Consequently, we had to augment our technical skills about dealing with radioactive wastes and learn how to be more effective in working with impacted communities. We learned the importance of being good listeners and being able to communicate complex risk issues in a relevant, understandable, credible, and nontechnical way.

I have worked on many water projects over the years. Almost all of them involved working on very complex and technical problems that impact people's lives. Thus, they inevitably had their social or political aspects, which I believe really makes water resources management so satisfying and enjoyable. It allows you to work on complex interdisciplinary problems that affect society and, at the same time, forces you to continue to grow intellectually, both as an engineer or scientist, but most important, as a person.

Following My Interests and Instincts

Laurel Saito

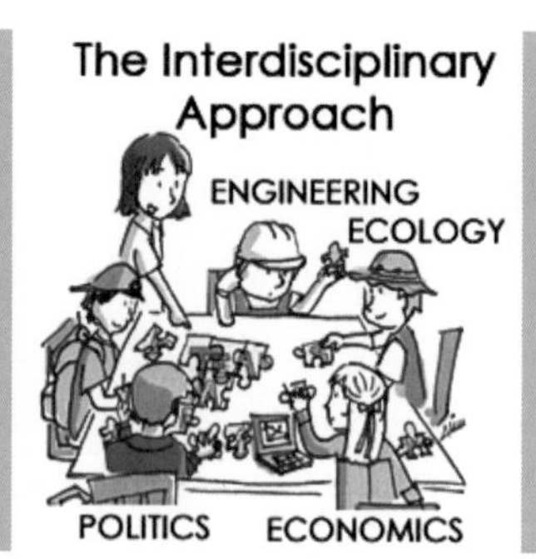

While working as a water resources engineer, I noticed that many of my supposedly engineering projects involved aspects of ecology, economics, politics, and other disciplines. I felt my engineering training had not prepared me well for these interdisciplinary aspects of water resources planning and management. Hence, when I decided to pursue research for my Ph.D., I wanted to make sure I would learn how to address interdisciplinary aspects of water projects. I chose to work on a project that addressed release policies at Shasta Reservoir aimed at improving downstream water temperature conditions for endangered Chinook salmon as well as generating hydropower. My research project was aimed at determining if proposed new reservoir operations would adversely affect the popular cold and warm water fisheries in Shasta Reservoir.

While working on this project, I ended up applying a couple of ecological modeling approaches linked to the output of a reservoir operations model. As an engineer, I had never heard of these ecological models, but an ecology professor on my graduate committee suggested I use them. They ended up being excellent ways of assessing how the reservoir operations could impact fish. This research experience made me wish there were a university course I could have taken to become aware of modeling from the perspective of other disciplines and to think more broadly about how we could apply multidisciplinary approaches to address complex water issues.

After graduating, I became a professor at the University of Nevada, Reno (UNR). I took that position with a strong desire to develop an interdisciplinary graduate program focusing on water resources planning and management and to pursue interdisciplinary water resources research. The National Science Foundation (NSF) has a competition for young faculty called CAREER, which requires faculty to propose related research and teaching initiatives that they will complete in a five-year period. I submitted a proposal to use interdisciplinary models to assess reservoir operations impacts on ecosystems and to develop a course I had envisioned as a Ph.D. student on interdisciplinary modeling. The NSF program manager suggested that I would have a better chance of getting funding if I submitted a regular research proposal. He liked the idea of the interdisciplinary modeling course and suggested I put together a proposal to fund a workshop to develop the course.

I returned to UNR and debated about applying for course development grants. I was hesitant, because I was coming up on my three-year review, and as an untenured faculty member, I knew that my department valued getting nationally competitive grants and having peer-reviewed publications, and that organizing and teaching this course would require a lot of time that was not clearly research-oriented. I was fortunate to have a mentoring committee at UNR, and I asked all three of them for advice. I got conflicting responses. One member thought this would be a good thing to do because it was NSF funding, even though it wasn't for research. Another thought I should not do this because I needed to spend time getting competitive research funding, so putting that much time into teaching at this stage of my career wasn't advisable. Ultimately, I considered the reasons I had for pursuing a career in academia and felt that even if doing this workshop resulted in me not achieving tenure, if I didn't do it, I would always wish that I had given it a try and wonder how it would have turned out. Thus, I decided to follow my instinct and interests. Otherwise I would have always wondered "what if."

I applied for and received an NSF seed grant. This provided the funding needed for me to meet with several people that I wanted to involve in the workshop, helped me write the proposal for holding the workshop, and also covered a deposit for the facility where I wanted to hold the workshop.

It worked. With the assistance of a graduate student, I convened the workshop at Granlibakken Conference Center in Lake Tahoe with 21 faculty and 23 graduate student participants from 13 different institutions. The faculty were experts in different disciplines related to water resources modeling. The workshop addressed topics related to interdisciplinary modeling such as issues of scale, integration, calibration, uncertainty, and error estimation. We created a wiki page to share course materials as part of a virtual textbook that could be a resource for instructors who wanted to teach this type of course at any institution of higher learning. Eventually, I ended up offering this course five times, always as an interinstitutional, interdisciplinary course. Many of the faculty participated multiple times in teaching the course. We wrote and published two peer-reviewed articles about the course and edited a special issue of a journal that featured articles from different faculty on topics related to the course. In addition, the relationships I developed with the faculty who participated in the course led to several funded research projects. And I was awarded tenure.

Recently, I decided to leave the academic life and work for The Nature Conservancy (TNC). Looking back, I am very glad that I decided to pursue the development of that interdisciplinary modeling course for many reasons. When I left UNR for TNC, I was very satisfied with my career in academia, and part of that satisfaction is because I succeeded by following my instincts. I pursued the things I wanted to do in academia and it worked. I met some great colleagues through the course and learned a tremendous amount. I now feel much more aware of the perspectives and contributions many disciplines offer.

Starting It Simple

Uri Shamir

"Sometimes pipes just need a patch."

Way back in 1977, consulting engineer Chuck Howard and I were occasional consultants to the Water Department in Calgary, Canada. At an afternoon tea break, Jim Bouck, the city water engineer, told me that the city of 450,000 people experienced some 1,500 pipe breaks a year and that the cost of fixing each break was in the order of $1,000.

Thus, water pipe breaks were costing the city about $1.5 million each year. When each break was discovered, the engineers were faced with the dilemma of whether to replace the whole pipe or just repair the break and wait for more breaks to occur before replacing the pipe. As the pipe aged, the number of breaks per year increased. This decision-making problem struck me as a classic least-cost problem everyone is exposed to in any elementary economics class. Obviously, the longer one waits to replace the pipe, the lower the present cost of that replacement will be. But on the other hand, the longer one waits for pipe replacement, the more money will be spent in fixing an increasing number of breaks. The sum of the two costs is a typical decreasing and then increasing total cost curve over time, which can be used to determine how long to wait before replacing the pipe. Until the time when the minimum value on that U-shaped total cost curve is reached, any breaks should be fixed. When a break occurs after that time, the entire pipe section should be replaced and a new cycle begun for that section. We used the city maintenance records to forecast the increasing number of breaks as the pipe ages, the costs of fixing them, the costs of pipe replacement, and the current discount rate to define these convex total cost curves and the time after which pipes should be replaced. I recall that Calgary adopted the method in practice, and then improved on it over time.

Being an academic and thinking this approach might be useful for other public water utilities, I and several others wrote and published a paper in the *Journal of the American Water Works Association* (JAWWA) describing this "analytic approach to scheduling pipe replacement" methodology along with some sample data. That paper generated a flurry of criticism on what we hadn't considered rather than what we had or what we did. Our critics claimed that we did not consider a whole slew of variables that affect the rate of breaks of a pipe, such as pipe material, internal pressure, external loads, depth of the trench, method of laying pipes, ground temperature, and more. We had used only historical data of actual breaks, so this was absolutely true. But all this attention led to a string of studies on the impacts of these other considerations and to refinements of the model. It also generated a veritable industry of practice and consulting on water distribution pipeline maintenance.

What did we learn? Perhaps it was to share novel ideas but be humble enough to learn from others. We also learned that sharing ideas can lead to more work and even better ideas.

Red River "Flood of the Century" through Human Eyes

Slobodan P. Simonovic

"The flood destroyed our buildings, but this town is still our home."

On my way to work on that spring day of 1997 (at that time I was a professor at the University of Manitoba, Winnipeg, in Manitoba, Canada), I witnessed heavy mechanization moving south; later, trucks full of soldiers and volunteers, and even later, school buses full of people being moved from the valley to higher locations. People from the Water Resources Branch, like Larry Whitney, emergency flood spokesman (who numerous times delivered lectures in my courses), Rick Bowering, head of the Water Resources Branch, and Doug McNeil from the city, became everyday guests in every Winnipeg home through a regular process of updating everyone about the Red River flood. The battle against the Red River was raging in the valley. Tremendous efforts to protect property and reduce damage were going on in parallel with the expansion of the water over the land. The "Red Sea" reached up to 37 km wide and covered 1,850 km^2 in Manitoba. On April 27 (three days before the peak flow in Winnipeg), my colleague and friend Prof. Wendy Dahlgrin took me in her small plane for a flight above southern Manitoba. Our flight route and altitude were under the control of the military. My stomach did not agree with the bumpy flight of a small plane. However, one picture remains in my mind. From the altitude we were flying at, all I could see was water. We flew from Winnipeg south to the Canada-US border and back. The river channel could be recognized only by the tops of the trees poking above the water level. The scene looked unreal. Farmhouses still above the water and townships protected with ring dikes looked like small islands in an ocean. The flood dislocated many people and caused considerable economic damage on both sides of the border.

Situated in the geographic center of North America, the Red River originates in North Dakota and flows north. The Red River basin covers 116,500 km^2 (exclusive of the Assiniboine River and its tributary, the Souris) of which nearly 103,600 km^2 are in the United States. The basin is remarkably flat. The elevation at Wahpeton, the origin of the Red River, is only 69 m above Lake Winnipeg, the end of the Red River. The basin is about 100 km across at its widest. The Red River floodplain has natural levees at points both on the main stem and on some tributaries. These levees (around 1.5 m high) have resulted from accumulated sediment deposit during past floods. Because of the flat terrain, when the river overflows these levees, the water can spread out over enormous distances without stopping or pooling, exacerbating flood conditions. During major floods, the entire valley becomes the floodplain.

On June 12, 1997, the governments of Canada and the United States requested the International Joint Commission (IJC) to examine and report on the causes and effects of damaging floods in the Red River Basin and to recommend ways to reduce and prevent harm from future flooding. The IJC is a binational Canadian–United States organization established in 1909 to assist the governments in managing waters shared by the two countries for the benefit of both. To assist its binational investigation of the Red River flood of 1997, the commission appointed an International Red River Basin Task Force. The task force, comprised of members from a variety of backgrounds in public policy and water resources management, was to provide advice to the commission. The governments asked the commission to examine a full range of management options, including structural measures (such as building design and construction, basin storage, and ring dikes) and nonstructural measures (such as floodplain management, flood forecasting, emergency preparedness, and response), and to identify opportunities for enhancement in preparedness and response that could improve flood management in the future. I was appointed to serve on the task force together with four more members from Canada and five members from the United States.

Working on the task force was an experience of a lifetime. I participated in many public hearings across the Canadian and US parts of the basin, literally meeting thousands of people affected by the flood. I had an opportunity to hear horror stories of those who lost everything, listen to the rage of people who felt left without assistance, and meet with those who worked hard to save their families and property from damage. This work brought me in touch with basin managers in Canada and United States. We had extensive meetings with representatives of all governments (local, provincial/state, and federal). I was part of many technical, social, and environmental studies commissioned by the IJC. For the first time in my professional life I got an opportunity to understand the full extent of the impact my work had on people, the environment, and society in general. The task force prepared a December 1997 interim report that cautioned against complacency and made 40 recommendations for better flood preparedness in the short term. We submitted our final report to the IJC in July 2000.

One part of the experience I gained was related to human behavior and disasters. The IJC organized a set of public hearings immediately after the flood (in the summer of 1997) to obtain detailed insights into the scale of the disaster and the experiences of those affected, to identify the scope for providing interim recommendations (for use during the next flooding season), and to then create a final set of recommendations. One stop on our public hearings tour was the small city of Morris, located south of Winnipeg, approximately halfway between Winnipeg and the US border. The residents of Morris and farmers from the surrounding area shared their experience during the public hearings. Being one of the technical members of the task force, I had a role to discuss various existing measures (both structural and nonstructural) and initiate a dialogue about future measures that may increase the level of protection and preparedness of the region for future flood disasters. At the end of my presentation and the following discussion, I was approached by the mayor of the city of Morris. This middle-aged farmer, very smart and involved, enthusiastically supported my way of thinking about flooding in this region. He openly stood up and after a small congratulations publicly stated, “You are the expert, tell us what to do and how, and we will implement that. . .Your expertise is what we need, together with the support from all levels of government." I was very encouraged after this support, and my enthusiasm for helping people was at a very high level.

The work of the IJC task force and many professionals over the next three years resulted in many ideas and proposals for future flood management. With a group of collaborators, I developed an original procedure based on the multiobjective approach capable of integrating quantitative and qualitative information and aimed at assisting the selection process. This tool was used extensively with every new option that was considered and served us well for generating recommendations for the final report. The task force's final report drew together all the findings collected over the three years and made recommendations on policy, operations, and research issues. One of the major findings was that floods of 1997 magnitude and higher will be experienced in the basin in the future.

The IJC used the final report as the basis for public hearings in the basin before the submission of its report to the governments of Canada and the United States. We were in the city of Morris again, three years after the flood and after the community had fully recovered, to present and discuss our final report. The mayor of the city had not changed. Many ideas were accepted, and there was general support for the recommendations laid down in the final report. Somewhere in the middle of our meeting, the mayor took part in the discussion and passionately started to criticize the tone of the final report, "We understand the physical conditions of the basin and trust the technical work you have done, but you were maybe too strong in saying that we will be facing similar floods in the future. . . . These statements may affect the growth of our community . . . people may decide to move to other safer regions . . . we know how to manage disasters."

I was sitting in the car driving to our next stop for public hearings and thinking about the strength of our human survival instincts. Three years was sufficient time to forget the hardship and look at the past experience through different eyes. My thoughts were going in the direction of how we could incorporate these human dynamics into our work. Is there a way to include the behavior of the mayor of Morris in the decision-making models when trade-offs including human hardship, economics, present and future risks, and other issues are being assessed?

A Computer Program Named Randy Raban

Slobodan P. Simonovic

"Wow! This program makes decisions just like our senior engineer!"

Randy Raban was one of the leading engineers in Manitoba Hydro (Manitoba, Canada). He was a big man with a yellowish beard, strong face, and a very clear, commanding voice. These characteristics put many people communicating with him on the "listening" side. Randy was a prairie boy, and I remember him telling me stories about his family in rural Manitoba, where he witnessed early days of electricity when the first street lamps were mostly used as shooting targets.

When Randy and I started working together, he was a leading expert in managing operations of Manitoba Hydro power system. He held the position of energy and reservoir scheduling engineer.

Nearly all of the electricity Manitoba Hydro produced was clean, renewable power generated at 15 hydroelectric generating stations on the Saskatchewan, Winnipeg, Burntwood, Laurie, and Nelson rivers. The utility also operates two thermal generating stations (Brandon and Selkirk) and four small remote diesel generating stations (Brochet, Lac Brochet, Shamattawa, and Tadoule Lake). In 2016, their total generating capability was 5,690 MW. Today, Manitoba Hydro is a major energy utility providing the lowest electricity rates in Canada.

Most of the Manitoba Hydro generating capacity is located in northern Manitoba and takes advantage of the very large natural storage of Lake Winnipeg. Lake Winnipeg's regulation system allows the utility to control the outflows into the Nelson River where 70% of generating capacity is located. The regulation system is an extensive engineered system of channels and structures that allows 50% more water to flow out of the lake than would otherwise flow out naturally. This increased outflow capacity helps reduce flooding around Lake Winnipeg and helps optimize power generation along the Nelson River.

The Jenpeg Generating Station on the upper arm of the Nelson River is one of the key elements in the successful development of the hydroelectric potential of northern Manitoba. In addition to generating electricity, Jenpeg's powerhouse and spillway structures are used to control and regulate the outflow waters of Lake Winnipeg, which in turn is used as a reservoir to store water to ensure enough water is available to run the northern generating stations on the Nelson River. During freeze-up, outflow from Lake Winnipeg is temporarily reduced, typically for a week or two, to permit formation of a smooth ice cover that improves the Nelson River west channel capacity. This ultimately increases the total Lake Winnipeg outflow capability over the winter. After this ice cover is formed, maximum discharge operations can be resumed if required. Operations of Jenpeg station during the freeze-up period attempt to reduce the impact on local resource users. Monitoring of ice conditions upstream and downstream occurs on a daily basis during the freeze-up. For many years, Randy was the expert supervising operations during the freeze-up period. The experience he obtained during many years of freeze-up supervision was unique.

My interest at that time was focused on expert systems for water resources management. An expert system is defined as the embodiment within a computer of expert knowledge in a form that the system can offer intelligent advice on, or implement an intelligent decision about, some issue. Expert systems may function as an assistant or a partner to a human expert. The simplest way of encoding knowledge is through a set of rules that define conditions and the corresponding actions to be taken if these conditions are met.

For the Manitoba Hydro system, the traditional techniques of analysis to aid in decision making were not capable of analyzing the system with all of its interdependences and growing complexities in the timely manner required for operations. To improve and speed up the analysis required by the reservoir energy scheduling engineer, Manitoba Hydro relies on the Energy Management and Maintenance Analysis (EMMA) optimization model. The objective functions maximized by EMMA are the revenues from sales of energy and the value of water left in storage at the end of some time horizon, less the costs attributed to energy production and imports, water releases, and maintenance. EMMA requires massive input preparation and can produce a number of reports. Expert system technology was considered to help in pre- and postprocessing model inputs and results. Developing the Jenpeg freeze-up expert system was to be a learning experience on how to build more complex expert system that will be working with the EMMA model.

The plan was to "put Randy Raban into a computer box"—not an easy task. His knowledge and experience in running the operations of the Jenpeg Generating Station during the freeze-up were the main target, and work proceeded through intensive interviews and coding of the knowledge base. Because of the nature of the knowledge acquisition process (i.e., human interactions), many difficulties were encountered. At the start of development, my team (including myself and two graduate students) was unfamiliar with the problem domain; therefore, initial questions and discussions with Randy were quite general. As a result, he was not focused on the freeze-up problem and provided more general, and not very useful, comments on his expertise. The development team was overloaded with information during initial encounters with Randy. The approach then taken was to create rules from the information the team thought was important or understandable at the time. Randy revealed more of his actual expertise through reviewing these rules. More specific and expert problem-solving knowledge was uncovered only after the development team made an initial attempt to establish rules. Thus, the initial prototype reflected the understanding and problem-solving ability of the development team.

The process lasted for some time and generated very useful experience in developing expert systems for hydropower domain. I learned that the development team must include persistent and patient people who will ask questions, understand the problem, and continually demonstrate what has been learned from previous interviews. It was clear that there was an advantage in starting with a small, simple, preferably structured problem to gain experience. The whole process of an expert system development depends on a cooperative, enthusiastic expert who is aware of the large time commitment required. Acquiring knowledge is not a linear process but rather an iterative process. Knowledge is elicited piece by piece. For water resources, engineering problems' rules are often the most natural and understandable form of knowledge presentation.

In early stages, both Randy and I were a little discouraged by the lack of significant progress. All the difficulties experienced were attributed to my lack of understanding of the problem. It was only by developing the prototype expert systems that a more suitable and achievable focus was identified. Randy was "in the box" and was happy there. The ground for a more complex EMMA assistant was prepared, and the freeze-up expert system was able to reproduce the experience Randy had accumulated over years.

As Is Often Said, "Rome Was Not Built in a Day"

Jery Stedinger

"Engineering guidelines are not written in a day."

I finished my Ph.D. at Harvard in 1977. While there, I had not thought much about estimating flood frequencies. But I recall hearing lectures on that subject from my professors Myron B. Fiering and Harold Thomas, and from visitors including Nick Matalas and Jim Wallis. They were developing new approaches to flood frequency analysis that they hoped would be adopted around the world, and by the US National Weather Service for precipitation frequency analysis.

But the publication of *Guidelines for Flood Frequency Analysis*, commonly known as US Bulletin 17/17A, and later 17B in 1982, sent US flood frequency analysis in another direction from the statistical methods I heard discussed at Harvard. Once I settled into my new job as a professor at Cornell University, I began to explore the efficiency of the methods outlined in those guidelines. Throughout the 1980s my students and I explored how historical flood frequency information could be used to improve flood frequency analyses. Moreover, it turned out that there were lots of such historical evidence around when one started looking for it:

newspaper accounts, historical letters and documents, high-water marks on buildings or in caves and alcoves, and even physical evidence such as slack-water deposits and scars on trees (such physical evidence is called paleoflood data).

Along with many other hydrologists, I continued to work on improving methods for using historical flood data, paleoflood data, and more generally, censored data wherein a flood flow value is known to be above or below a bound or in an interval. Slowly but steadily these methods were picked up and used in the United States and abroad. But Bulletin 17B did not change. Using the methods proposed in Bulletin 17B, flood frequency estimates were not nearly as good and sometimes made almost no use of historical information.

My graduate students were getting jobs in government agencies that carried out flood frequency analyses, and they and I, along with many other hydrologists, pressed on in seeking improved methods for using historical flood data. In 2000, two of our informal group of hydrologists (Tim Cohn and Bill Lane) developed an estimation method called the Expected Moments Algorithm (EMA). It was a direct extension of the estimation method published in Bulletin 17B, which employed improved methods of including historical and censored data. After Monte Carlo studies illustrated the improvements provided by EMA, we thought we were ready to take up the task of rewriting Bulletin 17B.

Thus, in 2005, the Hydrologic Frequency Analysis Work Group (HFAWG) met to consider the generation of a revision to Bulletin 17B employing EMA and the new statistical machinery that went with it. HFAWG had historically been a government interagency committee that developed Bulletin 17-17A-17B, among other bulletins. In 2005, nongovernmental scientists could be members if they were sponsored by an appropriate organization. Martin Becker, in particular, had joined the committee and had, for a time, served as its chair. He was a real estate developer who had developed a skepticism of government flood risk studies. In fact, in 2005, I joined with him to contest a US Army Corps of Engineers (USACE) report on a stream in Southern California.

An interesting thing about the HFAWG in 2005 was that three previous members had just retired; two were replaced with my previous graduate students. I also became active in the HFAWG as a nonfederal member. When HFAWG met in 2005, I was on sabbatical with the USACE. During a previous sabbatical in 1983–1984, I had been at the National Center of the US Geological Survey (USGS) working on Bulletin 17B issues. In 2010, I would return for a sabbatical at USGS to continue work on studies to support the adoption of EMA in Bulletin 17C.

By taking my academic sabbaticals with US federal agencies, I developed strong friendships and working relationships with engineers and scientists in those agencies. Making advances in the application of new scientific methods, and then getting them accepted and implemented within governmental agencies, required working with people and their organizations, as well as having ideas that work. Having my former graduate students in key positions in those agencies helped, too.

So, in 2005, the HFAWG believed it was ready to write a new set of guidelines. We believed that those new guidelines should be based on EMA. So why were the final guidelines only adopted formally in 2018, 13 years later? We thought we could update Bulletin 17B and have it approved in less than five years, but writing guidelines that would reliably work anywhere in the United States was a bigger task than we realized. Martin Becker and other HFAWG members wanted to see how the new methods worked on the "troublesome" records that had been investigated when developing Bulletin 17B in 1982. There was no agreement on how to do that. A separate data group was formed, drawing on committee members who were not EMA developers, to prescribe how the methods would be tested, but their prescription produced results that were difficult to interpret. Still, it became clear that when the approach was used with records from arid areas that contained many zero—or almost zero—flows, the application of EMA did not work well. Unusually small floods in many arid-area records can result in a major distortion of the estimated frequency of large floods. This was an unacceptable situation.

Hence, we had to make a major detour into statistical science. That was a big effort largely carried out by Tim Cohn (USGS) and my research group at Cornell, particularly Jonathan Lamontagne. Finally, we came up with a modification of EMA that was efficient, robust, and generally worked as well as or better than the Bulletin 17B procedures. The approach was peer-reviewed and published in *Water Resources Research.*

So, we did it. A draft of Bulletin 17C eventually was approved by HFAWG, the subcommittee to which it reports, and the Advisory Committee on Water Data, which was at the next level higher. There was a public comment period. Then, we had to respond to these comments and edit the draft guidelines, a task led by John England (US Bureau of Reclamation and later the US Corp of Engineers). Finally, in 2018, it all came together. And as much as our critics, including Martin Becker, frustrated us with their many challenges, questions, and lack of faith in our simulation experiments, they were really right; those of us with Ph.D. degrees were not as smart as we thought we were. But in the end, after many engineers and hydrologists pitched in when needed, we were all proud of the new national guidelines. A good number of people had participated in the effort, starting back in the mid-1980s, then more intensively from 2005 to 2018 Finally, after three decades of serious effort, Bulletin 17C became a reality. Did it have to be so difficult?

Cleaning Up Pollution and . . .Wait, What, Where?

Cristiane Queiroz Surbeck

"I'm in prison…
investigating pollution."

As I drove my little red coupe slowly down the empty street lined with rundown warehouse-like buildings, I heard a school bell. Suddenly I was surrounded by at least 100 men in blue jeans and white t-shirts walking from one building to the next. They walked on the street, forcing me to stop the car and wait as they passed by. They smiled when they saw me, but I had been instructed to look straight ahead.

It was the fourth month of my first job after college. I had landed my dream job at an environmental engineering consulting firm as a remediation engineer. I was learning how to investigate toxic chemical leaks and apply technologies to clean up soil and groundwater. My main motivation going into environmental engineering had been to clean up pollution. And that I did. But looking back at those early professional years, what stands out are the "wait, what?" moments when I first realized things that had never occurred to me before.

I could not have imagined during my job search that I would be sitting in my car inside the heavily gated grounds of a minimum-security prison while inmates walked from the cafeteria to the gym. When I got my degree in civil engineering, I envisioned alternating between a nice office and polluted job sites. I did get the office, but the polluted job sites turned out to be more interesting than generic industrial facilities.

How did I find myself working in this prison? Being a minimum-security facility, it functioned a lot like a camp; its residents lived in dormitories, ate in a dining hall (the mess hall), and mostly went about their day on a schedule of classes and exercise time. The classes consisted of on-the-job training to prepare inmates to join the workforce after they were released. Some of those classes were in auto repair, farming, dry cleaning, and even treating drinking water. Many of the skills taught used the chemical perchloroethylene, also known as PCE or perc, a common degreasing agent. As the description implies, it removes grease from anything from auto parts to dry-cleaned clothes. Over decades of use in the job training classes, PCE, a cancer-causing chemical, had been improperly dumped on the ground and down manholes instead of disposed of in steel drums and sent to a hazardous waste landfill. Over time, the liquid percolated into the soil, contaminating the facility's groundwater supply several dozen feet below ground. My firm had been hired to investigate the sources of the pollution and provide design and construction services necessary for the prison to treat its drinking water and distribute it to the taps throughout the facility. On the day of the red car incident, I, along with a contractor, had the not so glamorous job of opening sewer manholes, inserting a video camera on tracks into the sewer pipes, filming the inside of the pipe as far as the cable could go, and watching live video of miles of wastewater, toilet paper, and surprised cockroaches.

So why was I driving around in my red coupe inside the prison that day? A combination of necessity and miscommunication. First, all the company pickup trucks were taken that day. It was a hassle to rent a car. And I had to drive inside the facility because I had to locate manholes where PCE might have been dumped. I hadn't been briefed on all the security rules, and I didn't know I couldn't drive a personal vehicle inside the prison. When I showed up for the job that early morning, the prison guards at the entrance reluctantly let me use my car. The prison didn't like it, and neither did my company. That little red car driven by a non-officer woman drew too many inmates' attention. Some inmates found reasons to ask me random questions, like, "Have you seen Shorty? Shorty is the guy with SHORTY tattooed on the front of his neck." This made for awkward situations. Both the prisoners and we contractors were under strict instructions not to interact with one another. After this incident, an unarmed officer was assigned to stay with the contractor crew and me at all times. The officer was unarmed because it was a numbers game: the likelihood that an inmate would steal a gun from an officer was higher than the likelihood that we'd actually be attacked. Armed officers were up in towers overlooking the facility.

I spent about three weeks working on that particular project, and the overall remediation program there is still going on 20 years later. I had many other projects that I worked on for years, all of them with their little adventures and "wait, what?" moments, such as the one with the rattlesnake (it didn't attack us), the one when I forgot that water and electricity don't mix (no one got hurt), and the one by the beach when two Olympic beach volleyball medalists were playing (I had been captain of my high school volleyball team, so seeing these players was a big deal).

At the prison, Shorty did eventually come by to introduce himself. He was easy to recognize, and not just because he was short. He was all muscle. He said that if any of the guys were bothering me, I could call on him for help. I said an automatic okay, afraid to be caught talking to an inmate. And then after a "wait, what?" moment, I said a cautious—and sincere—thanks, before grabbing the crowbar to open the next manhole.

Engineering Your Own Education

Desiree Tullos

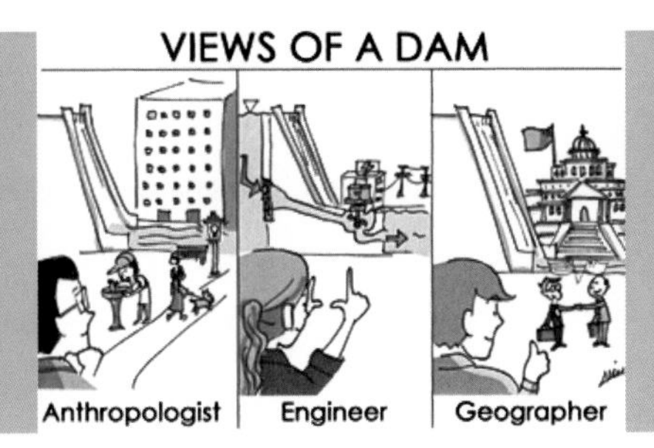

There have been many moments in my "education" as an engineer, the vast majority of which happened outside of undergraduate and graduate training. Broadly, those many moments of humility, curiosity, and panic taught me the life lessons that motivate me to get out of bed every day. In particular, I have come to appreciate that engineering alone cannot solve society's problems.

As with life, those educational moments happened when I was outside of my comfort zone and forced to reflect on others' disciplinary and cultural traditions. I have had the great opportunity to work in a dizzying diversity of places, from the hot, crowded, polluted, noisy cities of southeast Asia, to the stunningly beautiful but quietly tragic mountains of central China and the Indian Himalayas, to the politically and historically mystifying issues around water in the Middle East. Despite the intrigue and fascination of working in places so different from my home, I admit that the most complex problems that I have worked on in water resources engineering are those in the United States. From the passionate hornet's nest of issues in California's Sacramento–San Joaquin system, to the polite but technically challenging issues of designing effective fish passage and water rights in Oregon's Willamette basin, it is tempting to think all of the world's juiciest water problems are overseas. The reality is that water touches all human, animal, and plant life in the world, and we need to act both locally and globally to develop ourselves and to engineer solutions to water crises facing the world today.

With that said, I now summarize some lessons on engineering that I learned during an eight-year study on the impact of hydropower development on the isolated ecosystems and people of central China. The project launched at the start of my career when I was a very young and very green assistant professor, and it laid an important foundation for a lifetime of curiosity and humility about engineering resources as essential as water and energy.

The project was a National Science Foundation–funded investigation of dams as agents of change. I was the lead PI of an all-star team that included Brian Tilt (anthropologist emphasizing environmental justice in China), Phil Brown (economist studying environmental economics in China), Darrin Magee (environmental scientist whose research is still an enigma to me), and Aaron Wolf (a geographer whose specialty is in transboundary conflict resolution), along with a cadre of *really* bright students. We were a motley crew, and working with them deeply transformed my view of the world and my place in it.

Engineering education is a funny thing. We teach students how to decompose problems into a set of given, required, and assumed conditions, and ask them to apply an existing model or standard to arrive at a single right answer. This paradigm made sense as an undergrad when we were designing bridge trusses, but I have never encountered a single problem in river engineering that conforms to this approach. Ironically, I still teach this approach to students because I do think there is value in systematically articulating all the components of a complex system. However, even in the simplest of examples, for example when you arrive at a grain size needed for a "stable" channel or the ballast needed for an engineered log jam, you have only completed the first of many steps in a design. There are countless uncertainties (e.g., how confident are you in the design discharge as the catchment and climate change or in the use of a widely used criterion for incipient motion of sediment) and social dimensions (e.g., how hard will it be to get a permit for this project? How will the public respond to a large accumulation of wood along a recreational reach of river?) to all engineering problems that make our own training paradigm somewhat laughable.

I got my first glimpse of the arrogance and oversimplification of engineering while writing our first proposal to the National Science Foundation. Our team sat down at a big table to start brainstorming the framework we would ultimately propose for dams as agents of change. Confidently, I started the conversation by saying, "Well, dams are fundamentally driven by hydrology and hydraulics, so that should be the foundation of our framework." My colleagues laughed. One responded by saying, "Actually, dams are motivated by people and their need for energy and water, so it has to begin with the social dimension of dams." Another countered, "A dam would never be built if the geopolitical climate didn't align to create the institutions that establish the laws and funding for design and construction of the dam."

Since that moment in 2006, I have come to appreciate that designing a dam is by far the simplest aspect of dam development. The laws and regulations governing where dams can be located, which impacts can and must be mitigated, and how the structure should be operated are infinitely more complex than the calculations of reservoir storage, spillway discharge, and downstream flood stage. And the social dimensions of water infrastructure will likely always be a mystery to me. Those social dimensions define why we build dams that make absolutely no economic or practical sense, why the public reliably fails to take responsibility for our own flood risk, and why we insist on maintaining infrastructure that is unsafe and no longer providing its intended benefits.

The reality is that all engineering problems facing the world today are plagued by this complexity, where complexity refers to a system with a large suite of poorly understood interactions between elements in the system. Thus, no single discipline or perspective fundamentally owns the complex problems that engineers are attempting to solve today. Engineers are problem solvers, but we need to engage the expertise of others to fully understand the diversity and implications of potential solutions. The future depends on engineers acknowledging these interactions and being comfortable working within and outside of the given-required-assumed-solution paradigm.

In summary, I encourage students to take their engineering education seriously, as there is great value in learning the physics of fluids or soils and in developing the ability to coherently structure a problem to be solved. However, I also emphasize that this knowledge is not enough to do the difficult work of sustainable engineering of the world's natural and human resources. Whether you work in engineering solutions for energy or water or technology or ecosystems, commit yourself to being curious and humble; to recognizing that you can learn from anyone, including those with little or no formal education and those with very different perspectives from yours; and to moving outside of your comfort zone by pushing yourself to experience new places and disciplines. Spend time out in the field you are working in, collecting your own data, talking to locals, and riding the wild rapids of life and big-river hydraulics.

I leave you with a quote from Gilbert White, a personal hero of mine for his work on sustainable floodplain management: "The good life, like the balance of all the complex elements of a river valley, is founded upon friendly adjustment. . .It embraces confidence in fellowship, tolerance in outlook, humility in service, and a constant search for the truth."

Effective Communication

Desiree Tullos

"He's a great engineer, but I just don't understand him."

When I ask practicing engineers what the most critical skills are for graduating students, I consistently hear the same three things: GIS and CADD modeling, programming, and effective communication. The first time I heard that effective communication was a priority, I was a bit unsure of what to do with that information. None of us received training on this as undergrads. How would we integrate that into an ABET-based degree? And what does effective communication even mean?

I have had a few generous teachers mentor me in the importance of and strategies for communicating effectively. My husband has been one of my most patient teachers. He continues to teach me the value of directly expressing my ideas and concerns, and that listening is about more than just hearing words.

Another of my teachers in this regard is Aaron Wolf. If you have ever met Aaron, you will understand why he is such a great teacher. He is 100% honest, sincere, and joyful, yet very persuasive and convincing. I have witnessed him soften even the most hardened adversary. It is nothing short of magic, a magic that I have been trying to learn since we started working together in China in 2007. For our first meeting with our local project partners, I prepared in the way that I believe is most effective as a Type A engineer. There was a very neatly detailed agenda, an explicit outline of partner contributions with deadlines, and mechanisms for exchanging data and products. As the project's principal investigator, I thought I was doing everyone a favor by doing the work to articulate a draft of the details on how we would work together. The response of our Chinese colleagues? Absolutely not. They would not agree to any of it. Following some confusing and vague attempts to gain clarity of what I thought we had already committed to in the proposal process, I began to feel the panic that the project was spiraling toward complete failure.

But in that moment, Aaron managed to gently redirect the conversation. It was so subtle. A loosening of a tie. A suggestion for a tea break. Informal discussion about each other's families. An undetected transition back to the future of rivers and hydropower in China. A question about trajectories of research and management. The alignment of interests for the research team. And somehow, through this sequence of unnoticed redirection and manipulation of the conversation, Aaron's magic transformed the group dynamics and ultimately the outcomes of the project. Our colleagues agreed to everything on my original outline of contributions, and more.

There are a lot of lessons embedded in this and other conversations that Aaron has led. In loosening his tie, Aaron was signaling to our colleagues that the meeting was not quite as formal as it appeared, and we should all soften our positions a bit. By talking about our families and the future of rivers, he was helping us identify that we all shared common values. In effective communication, one needs to be willing to build relationships and trust, which takes time, feels inefficient, and requires an enormous amount of patience and control of nonverbal cues. However, these investments in relationships can result in far more productive conversations and are more likely to produce desired outcomes. In helping to resolve subsequent conflicts, he has also taught me that the person with greater power in a relationship needs to be the more patient and humble listener, and that I do not need to try to correct or defend myself from every assault launched my way.

More than any other skill, I encourage young engineers to develop a strong writing practice, verbal and nonverbal communication skills, and an ability to facilitate discussions among individuals having conflicting goals and different opinions. In your professional writing practice, learn the rules of grammar and apply them unfailingly. Use as few words as possible to communicate about your work. Take care in organizing your ideas so that others are unlikely to misunderstand them. In verbal communication, gain some awareness of your poker face. Ask a friend to video you while you debate a passionate topic with another friend. Practice putting yourself in others' shoes so you can understand what values and fears underlie what seems like a completely irrational position to take. Recognize that everyone has an interesting story or insight to share. Take a course in conflict resolution and mediation. And before responding to any insults, have a beer and wait 24 hours before sending that email, or in person, take a long, deep breath. Or six. The result will be a much higher level of productivity and satisfaction, and a much lower blood pressure. Speak up for those around you who have less opportunity to speak for themselves. Listen to some podcasts or read books to develop an awareness of your unconscious biases and how to create inclusion in your communications and interactions with those around you. Whether you are on track to become a CEO or a field technician, these skills will benefit you individually and your engineering practice as well.

Broadening your engineering practice will require recognizing that your view might not be the only valid way of seeing the world or a problem, and to develop this practice and perspective, you have to be an effective communicator, a skill in which engineers are notoriously weak. Effectively communicating goes beyond grammar and well-designed PowerPoint slide decks, though these remain weaknesses in the engineering education. Learn to really listen—to everyone. Learn to choose words carefully when you speak. Be aware of your own biases and communication practices that are counterproductive. When we look beyond the confines of what is given, assumed, and required, we see that engineering solutions to the world's problems is such a rich, satisfying, humbling, and challenging task.

Part VI

Adventures in International Organizations

The Dublin Principles on Water and the Environment

Jerome Delli Priscoli

In 1992, an International Conference on Water and the Environment was held in Dublin, Ireland. The output from this conference was a declaration regarding water that was presented to the United Nations Conference on Environment and Development (UNCED) that was held in Rio de Janeiro later that year. The Rio conference, which came to be known as the "Earth Summit," was attended by 118 heads of government and was a major turning point in bringing the issues of sustainability and sustainable development onto the international political stage. The inclusion of the Dublin Principles in the conference debate helped to highlight the importance of water as a resource for environmental protection and human development. The Dublin Principles remain the standard for consideration of the issues surrounding water resource use and protection. The principles are listed below:

- **Principle No. 1:** Fresh water is a finite and vulnerable resource, essential to sustain life, development and the environment.
- **Principle No. 2:** Water development and management should be based on a participatory approach, involving users, planners and policy-makers at all levels.
- **Principle No. 3:** Women play a central part in the provision, management and safeguarding of water.
- **Principle No. 4:** Water has an economic value in all its competing uses and should be recognized as an economic good.

Before the Rio conference, I was working at the World Bank with the senior water advisor on the World Bank's first water policy. Many in the water community from international organizations feared that reports stemming from Rio would leave water out; that it would be lost amid other environmental concerns. Hence the purpose of the water conference in Dublin was to discuss and articulate the world water community's concerns. I do not think anyone anticipated the powerful impact the output of this conference, the Dublin Principles, would have over the tumultuous next decade of political-diplomatic and professional debates on water.

At the last minute I was asked, by the UN and the World Bank, to design a process involving hundreds of experts at Dublin to generate water proposals for the upcoming Rio conference. The World Bank arrived with its proposals, as did others. Suspicions of hidden agendas of the powerful were rampart.

I immediately suggested a radically different (for the time) format for discussions. We created a series of roundtables that facilitated discussions among participants. We engaged hundreds in discussions that lasted all day and into the evening. We came up with a series of statements. These statements we then posted on a large wall for the next morning. As participants entered the room that next morning, they were asked to check agree-disagree on a five-point scale under each proposal. The basis of the four Dublin principles readily emerged.

When the Dublin principles were brought to the official UN Rio conference, they were shunned, as they did not meet official international procedural standards. Significant objections were raised, but water did get mentioned in a chapter of the document reporting on the Rio conference. However, the consensus that was built in Dublin, and reflected in the four principles, was so strong that they held and were supported by the wide majority of water community regardless of what the official UN Rio report said. Consensus among the affinity groups of water professionals and NGOs overrode formal international organizations and changed the public diplomacy of water. The water community still holds to these principles, which have been adopted by those formal organizations that initially rejected them. They have been a basis for many subsequent water management initiatives. This has been a lesson in the power of consensus in public diplomacy.

The World Water Council (WWC) and World Water Forums (WWF)

Jerome Delli Priscoli

In the late 1990s, the world water community met in Marrakesh, thanks to funds from the World Bank, the King of Morocco, and others. This meeting, which I attended, was called the first World Water Forum (WWF). From its modest beginnings of several hundred water ministers, the World Water Council (WWC) and World Water Forums were born. The World Water Council was formed to find new ways of bringing technical and political leaders together around water policy.

The second forum was planned by the Dutch. This second forum astounded all of us—thousands (not just hundreds) wanted to attend from all sectors and parts of the world. Crown Prince William (now King of the Netherlands) took charge of this forum and has become one of the world's most important advocates for water.

The second forum produced the world water assessment process within the UN system. The World Water Assessment Programme produces periodic reports on the state of water resources and their management in the world. The third forum convened a high-level financing panel formed by the head of the International Monetary Fund, which resulted in policy advocacies that placed water on the agenda of a G-8 meeting. The Istanbul forum provided a safe meeting place for riparians on the Tigris and Euphrates to commit to significant new data exchanges. It also produced what is called the Istanbul Consensus: a document signed by mayors of cities around the world committing to adaptation measures for climate variability.

The WWFs have now grown to 20,000-person events that take place every three years. They are the largest and most diverse meetings of interested water groups worldwide. Members of the WWC now include hundreds of the world's major water organizations, including private, public, not-for-profit, advocacy, and other groups.

It is easy to criticize the large WWF events, and many do. Every three years we hear, "Look at all the money spent on these events; why not send it to the poor who need it?" From a perch on their international steering committees, I have watched the WWFs become a new type of public diplomacy instrument for water. They are not just meetings or big conferences. These events include heads of state, much of the world's royalty, hundreds of ministers of water and environment, hundreds of parliamentarians, hundreds of local officials, thousands of water experts, and advocacy groups—all observed and reported on by the world's media outlets.

Actually, the WWFs appear to have grown organically into a world platform for interests to form partnerships and to strategically plan. For example, from NGOs themselves we hear criticism: there are not enough poor people, not enough indigenous people, too many water companies, and more. But they also say, over and over, "Do not stop the WWFs;

they are the only place we can meet water decision-makers on such a scale." In fact, each forum attracts an alternative protest forum. The WWF attraction is so large it essentially provides a great platform for protests of all sorts.

We make great efforts to integrate the protesters into the WWFs. However, after talking with a world-renowned person in water protest movements at one WWF, I realized what should have been obvious: the alternative protest movement was essentially being provided an inexpensive world stage by the WWFs. Indeed, the WWF is a platform, an event, or a happening that is used to affect public diplomacy worldwide. Maybe someone can coin a good name in this new world of public diplomacy.

These WWFs started with a focus on ministerial declarations. Over time the importance of such declarations has diminished as the number of political actors attending the forums has expanded, and the forums are now producing declarations of parliamentarians, local authorities, city mayors, and various regional leaders.

The WWF outputs and the WWC have played a key role in making the world's political leaders and international organizations, such as the World Bank and the UN, aware of the importance of water and its management.

Changing the Terms of Discourse in World Water Debates

Jerome Delli Priscoli

We must work to raise the visibility of water management issues as well as water literacy worldwide. Public diplomacy and the media are central to such efforts. However, the way this is currently being done can be quite dangerous. To me, nowhere is such danger clearer than in public diplomacy debates on "water and war," and on "water and climate change."

Water and War

I have spent much time objecting to the water and war thesis, as its use in public diplomacy risks creating the conflicts it seeks to avoid. Our news has been filled with statements like “water is the next oil,” “water wars,” and “water as the liquid of the 21st century.” For example:

> “The only matter that could take Egypt to war again is water” (Anwar Saddat, the third President of Egypt, 1979).
>
> “The next war in the Middle East will be fought over water, not politics” (Boutros Boutros Ghali, sixth Secretary-General of the United Nations, 1985).
>
> “The wars of the next century will be about water” (Ismail Seageldin, Vice President of the World Bank, 1995).
>
> “Ever-increasing global demand for the scarce water resources that we have will almost certainly lead to future geopolitical conflicts” (Jeff Bingaman, US Senator, 2005).

But are these statements more misleading than illuminating? In 1995, the Israeli Defense Forces analyst responsible for long-term planning during the 1982 invasion of Lebanon noted,

> “Why go to war over water? For the price of one week’s fighting, you could build five desalination plants. No loss of life, no international pressure, and a reliable supply you don’t have to defend in hostile territory.”

Even some of the commentators, such as Kofi Annan, the seventh Secretary-General of the United Nations, seem to be changing. For example,

> "Fierce competition for fresh water may well become a source of conflict and wars in the future" (Kofi Annan, March 2001).

> "But the water problems of our world need not be only a cause of tension; they can also be a catalyst for cooperation. . . . If we work together, a secure and sustainable water future can be ours" (Kofi Annan, January 2002).

If one sees water as a matter of fighting over allocations of limited flows, as many in our security community now seem to think, it is easy to come to the conclusion that there could be water wars. However, the history of water resources management's role in the development of civilizations has been very different, especially at the macrosocial levels. That history has been one of learning (albeit often painfully) how to jointly create and share benefits through negotiating the multiple uses of water. Such efforts never stop as both supplies and demands and people's water use preferences change as climates and societies change.

Oregon State University data show that for more than 1,800 conflicting and cooperative water interactions over the last 50 years, only seven disputes involved violence. Incidences of violence are more frequent at individual or small community levels than at larger societal levels.

The water crisis is mainly one of distribution of water and reapportionment of water uses, not one of absolute scarcity. However, the water and security debate is driven by notions of scarcity. Although water may be used by parties in conflict, water is rarely the cause of war and large-scale social violence. However, such opinions dominate the security and water debates.

Consequently, the most salient public diplomacy aspects of water are often passed over in the debate. These diplomacy aspects are water's powerful historic roles in building social community; generating wealth through provision of preconditions of economic activities; convening adversaries and providing common language for joint and creative dialog; integrating, in a practical way, diverse interests and values; and providing a principal tool for preventive diplomacy and for building cultures of cooperation, if not peace.

Water and Climate Change: The Ethics of Public Diplomacy Debate

At the 2010 United Nations Climate Change Conference in Cancún, Mexico, an attempt was made to bring the climate change and water communities together. I was asked to introduce the workshop on water and climate change. I started by posing the question, "Is our public diplomacy raising people's anxiety without providing means (adaptive water actions) to deal with this anxiety?" Regardless of where one might stand on the debate over causes of climate variability, this question needs to be asked.

The major reasons repeatedly used in the talking points of international officials, such as the UN Secretary General or the head of the Intergovernmental Panel on Climate Change (IPCC), for why we should deal with climate change are water related. They include impacts of extreme hydrological events, increased frequency and intensity of droughts and floods, sea level rise, reduced water access and increased scarcity, water quality and health issues, and others. However, climate change debates seem to focus on long-term globalized mitigation measures, while appeals to the public on why we should be concerned focus on shorter-term extreme events that will occur no matter what long-term measures are implemented. This raises the important ethical public policy question: are we raising peoples' anxiety without providing means (adaptive water actions) to deal with this anxiety?

For example, consider the Indus River flood that occurred in late July and August 2010. That flood led to a humanitarian disaster considered to be one of the worst in Pakistan's history. It affected approximately 20 million people, destroyed homes, crops, and infrastructure, and left millions vulnerable to malnutrition and waterborne diseases. Estimates of the total number of people killed ranged from 1,200 to 2,200, while approximately 1.6 million houses were damaged or destroyed, leaving an estimated 14 million people without homes. It was not, as often touted, an unusual climate change event: it was closer to an event expected only once in 50 years on average. Post 2010, Indus flood calculations project that, had planned Indus River water storage been in place, most of the suffering and damage would not have occurred. Is this same dynamic of socioeconomic growth, along with decreasing measures to manage the vulnerabilities of that growth to water-related disasters, becoming true in other parts of the world that are also of great security interest? What does this say about the dialog between climate change, which is introducing more uncertainty to the public, and the water community whose business it is to help that public manage uncertainties often primarily with water investments?

Here is one prominent environmental nongovernmental organization's view. Advocating long-term climate change mitigation measures, the spokesman publicly noted: "If we do not do anything about climate change the people of Bangladesh will continue to be flooded. . .and. . .we can no longer engineer our way out of the crisis of climate change." Is this how to frame the debate?

The water manager's traditional role is to minimize risks and the costs of hazards to society by working at watershed levels using probabilistic models and methods. Their infrastructure designs are based on how extremes of floods and droughts are defined over time. Currently, global circulation models (GCMs) by and large do not yet help water managers deal with the very phenomena that are so prominent in public concerns about climate changes. They are being advocated for purposes they were not designed for. Information from GCMs does not offer adequate reliability for precipitation and runoff. Nevertheless, climate models leave water managers to contend with several GCMs generating hundreds of scenarios. This is juxtaposed with over 100 years of peer-reviewed analytical approaches to the risk and uncertainty of extreme events within the hydrological community.

The large uncertainties push decision-makers into clearly unaffordable options, even for the rich countries. Thus, we need to ask if the uncertainties introduced to the water policy debate are so large (and water managers in many parts of the world already use risk and return rates upward of 500 years for public investment purposes), is it rational or even ethical to be telling decision-makers to change? As general rapporteur for the Arab Forum in November 2012, I heard similar statements such as "stop forcing us to use GCMs; they are telling us to implement policies that will bankrupt us."

So are climate change and water debates indeed raising public anxiety while inadvertently denying adaptive means to cope with projected hydrologic events that will occur because of those changes? Does this raise questions about the ethics of adaptation versus mitigation?

Developing a World Bank Water Sector Strategy

Jerome Delli Priscoli

"Who are we working for again?

In the early 2000s, building on the World Bank (WB) water policy of the 1990s, the World Bank water teams made several attempts to pass a revised water sector strategy through the World Bank's board of directors. The new policy had been field tested over the previous three years. All attempts to get the board's approval failed, in large part by an ideology that strongly opposed funding large infrastructure and dams. Frustrated by these failures, the WB senior water advisor asked me to design and run a participatory evaluation on all continents of the proposed water sector strategy. I accepted, on the conditions that this evaluation be transparent, be run exactly the same way in all countries, include a cross section of key stakeholders (indigenous as well as external), and that the results would not be altered. Several weeks after completion, an elated WB water advisor called me and said, "You may be interested, the bank's board just approved the sector strategy, after all these previously failed attempts."

What happened? Stakeholder preferences from around the world, visually reflected in the distribution of simple dot voting on flip charts, produced a powerful message in stark terms. That visual captivated the WB board. It forced them to ask: "Who are we working for?" Indigenous NGOs strongly favored the sector strategy, whereas international NGOs based in Western capitals opposed it. The visual display of this was dramatic.

The realities underneath these results were not new and other factors were at play. However, the participatory evaluation was clearly catalytic. What had become a strong ideological coalition of Western donor nations fueling a specific ecological conditionality was overcome by the realities of clients' needs, clients for whom the international institution was supposed to be working. I was as surprised as anyone.

My takeaway from this experience follows:

> Transnational stakeholder participation can impact serious international policy. After all, once the World Bank writes policies or sector strategies, most of the other banks follow and so do the many donors.
>
> We must understand that our well-intended rhetoric, based on realities we experience in the rich world, often comes across to the poor as, "Do what we say, not what we did," or "Preserve, but do not use." All of these likely sustain, rather than transform, poverty. Our public diplomacy must address the question of how to use water to create platforms for growth while designing mitigation of costs to environment.
>
> Most religious faith traditions across civilizations use water as symbol of reconciliation, healing, and regeneration; we should remember that water debates mirror debates of social ethics.

Index

C

D

E

F

G

H

I

J

K

L

M

N

O

P

Q

R

S

T

U

Y

About the Authors

Thomas Ackermann, Ph.D., RWTH Aachen, Cornell University, started his career in hydraulics and water resources systems planning. As an engineer in a consulting firm, he developed infrastructure projects in Australia, both for construction companies and sponsors of various projects, and was responsible for international hydropower development activities. Since 2016, Thomas has been a professor of environmental science and hydraulic engineering at the Munich University of Applied Sciences. <thomas.ackermann@hm.edu>

Lily (Sanchez) Baldwin is a senior water engineer in the environmental technology unit at Chevron Energy Technology Company's Health, Environmental and Safety Department. Prior to Chevron, she worked at Lawrence Livermore National Laboratories and was a consultant prior to that. Lily has a passion for responsible water resource management while meeting business needs. She also enjoys singing in women's barbershop groups. <lilysbaldwin@gmail.com>

Emily Barbour, Ph.D., is a senior research scientist at the Commonwealth Scientific and Industrial Research Organisation (CSIRO) Australia. Emily's research focuses on the application of modeling tools to aid decision-making in water resource management and the assessment of trade-offs between multiple objectives. This involves the integration of hydrology, ecology, water management policy and operations, stakeholder engagement, and multiobjective optimization. Emily's work currently focuses on water security and poverty dynamics in South Asia. <Emily.Barbour@csiro.au>

Thinus Basson, Ph.D., P. Eng., was an executive director at BKS Consulting Engineers in South Africa, where he headed the Water and Power Division. After graduating at universities in South Africa and studies at the

University of California, Berkeley, he specialized in water resources planning and management. He has worked for several international agencies on some of the largest river systems in the world. <bassonthinus@gmail.com>

Mikhael Bolgov, Ph.D., Dr.Sc., is vice-director and head of the Surface Water Modeling Laboratory of the Institute for Water Problems of Russian Academy of Sciences. He is a specialist in the field of hydro-meteorological processes and phenomena and has developed hydrological forecasting methods given the nonstationary character of climatic change uncertainty. <bolgovmv@mail.ru> <bolgov@aqua.laser.ru>.

Ben Braga is a professor of civil and environmental engineering at the University of São Paulo. After obtaining his Ph.D. in the late 1970s from Stanford University, he joined the university to teach water resources systems modeling and create the environmental engineering major. After 20 years of teaching and consulting, he took a sabbatical at Colorado State University, where he became interested in other aspects of water management beyond mathematical modeling. From there on he served on the board of directors of the National Water Agency in Brasilia and later as Secretary of State for Sanitation and Water Resources in Sao Paulo State. He was president of the International Water Resources Association and is now finishing his second mandate as president of the World Water Council. <b.braga@worldwatercouncil.org>

Alexander Buber graduated from the Faculty of Mechanics and Mathematics of Moscow State University. As deputy director of "Soyuzvodproekt," he supervised the scientific and technical cooperation with institutes of the German Democratic Republic and the Hungarian People's Republic. Since 1990, he has been engaged in management of the water resources of the Russian river basins Yenisei, Angara, Volga, Amur, Kuban, and Don and in the operation of the reservoirs of the Volga-Kama Cascade. <buber49@yandex.ru>

Graeme Dandy, Ph.D., has more than 40 years of experience in teaching, researching, and consulting in the fields of optimization and artificial intelligence techniques applied to the planning and management of water resources and environmental systems. He is currently an emeritus professor in civil and environmental engineering at the University of Adelaide, Australia. He is also a founder and former director of Optimatics, a software and consulting company (now owned by Suez) that specializes in the optimum planning, design, and operations of water supply and wastewater systems. <graeme.dandy@adelaide.edu.au>

Ben Dysart Since obtaining engineering degrees from Vanderbilt University and Georgia Tech, Ben has had careers in academia, business, professional organizations, and government. At Clemson University he created and led Environmental Systems Engineering and Water Resources Management graduate programs. He served as president and chairman of the board of the National Wildlife Federation (NWF), the world's largest environmental advocacy and education NGO. He has held advisory positions in numerous governmental agencies and private companies including the World Bank, EPA, and US Army Corps of Engineers. He is a trustee and vice chair of the board of directors of the Rene Dubos Center for Human Environments and serves on the national board of trustees of Trout Unlimited. Ben's goal in life is to work toward making a difference for the good of all people and our natural environment. <dysart.ben@gmail.com>

David Ford, Ph.D., is an internationally recognized expert in water resources engineering, planning, and management. As president of his consulting firm, he and his firm have provided consulting services to local, state, and federal agencies throughout the United States and internationally. Ford has been a key advisor to the US Army Corps of Engineers and the California Department of Water Resources in development of flood risk reduction policy and plans for the State of California, including the Statewide Flood Management Planning Program, the Central Valley Flood Protection Plan, and the Urban Levee Design Criteria. His areas of expertise include management of complex, multi-agency projects involving surface water hydrologic analyses, fluvial hydraulics, flood risk assessment and real-time forecasting, flood warning, reservoir system operations, decision support analysis, water resource planning, and hydro-economics. <ford@ford-consulting.com>

Walter Grayman, Ph.D., P.E., took his first computer class while in high school 56 years ago but never became a very elegant programmer. He once wrote a 2,000+ card FORTRAN program which included just one small subroutine and only a handful of comment cards. His training at Carnegie Mellon University, MIT, and at a couple of large and small consulting firms prepared him well for his ongoing 35-year stint as an independent consulting engineer in the water resources field. This has provided a great opportunity to participate in a lot of very interesting projects, travel widely, and, most importantly, work with a wide range of fascinating and accomplished people. <wgrayman@gmail.com>

Neil Grigg fell in love with rivers in Alabama and figured out how to make them a paying job, as well as a passion. This involved many stops along the way, including the US Army, jobs in a consulting firm, several universities, the North Carolina state government, and positions in a number of other countries. <Neil.Grigg@colostate.edu>

Leon Hermans was educated as a policy analyst at Delft University of Technology. He worked for a few years at the water service of the FAO before returning to TU-Delft. He now enjoys teaching and research at the intersection of policy evaluation, water governance, and delta management at both TU-Delft and IHE-Delft. <L.M.Hermans@tudelft.nl>

Janusz Kindler, Ph.D., Dr.Hab., professor emeritus, and former dean of faculty of environmental engineering at Warsaw University of Technology, has engaged in teaching and research in integrated water resources policy, planning, management, and operation of water resource systems. He has held positions in the Polish National Water Resources Council, Polish Academy of Sciences, International Council of Scientific Unions, World

Bank, and International Institute for Applied Systems Analysis. He continues to consult on various water resources and environmental projects at home and abroad. <jkindler@is.pw.edu.pl>

Stacy Langsdale, Ph.D., P.E., is a water resources collaboration expert with the Institute for Water Resources and US Army Corps of Engineers, which she joined after completing her doctorate. Currently, she teaches engineers and technical experts how to communicate the risks associated with flood risk management infrastructure (dams and levees) and environmental hazards on former military sites. <Stacy.M.Langsdale@usace.army.mil>

Deborah (Hollister) Lee is director of the Great Lakes Environmental Research Laboratory (GLERL) in the National Oceanic and Atmospheric Administration's Office of Oceanic and Atmospheric Research. Prior to GLERL, she worked as the chief of water management for the Great Lakes and Ohio River Division of the US Army Corps of Engineers. Debbie's passion is to find solutions that both protect the environment and enhance the economy. She also enjoys traveling to new destinations on holidays with her family. <deborah.lee@noaa.gov>

Leif Lillehammer has worked both as a scientist in freshwater ecology and as a water resources practitioner. He graduated from the University of Oslo, and in the early years also worked at the Zoological Museum in Oslo. Over the past 20 years, water resources projects have taken him to multiple parts of the world, where he has also frequently collaborated with academia, including Cornell University. <leif.birger.lillehammer@multiconsult.no>

Angela Liu earned her bachelor's of science degree in environmental engineering from Cornell University and her MD degree from Case Western Reserve University. She is currently a medical doctor in Baltimore, Maryland. She loves to draw and has enjoyed this opportunity to reconnect with her environmental engineering roots. <al488@cornell.edu>

Hugo A. Loaiciga, Ph.D., P.E., D.WRE, has been in the education and research profession for many years. He served as the water commissioner for the City of Santa Barbara, California, for six years before joining the University of California at Santa Barbara in 1988. There, he teaches and carries out research in planning, design, and analysis of water resources systems and the theory and computational aspects of surface and ground water hydrology. He does not cease to be amazed by the disparity between plans and outcomes. <hugo@geog.ucsb.edu>

Alexander V. Lotov, Ph.D., Dr.Hab., graduated from Moscow Institute (University) for Physics and Tech. He has been affiliated with the Computing Centre of Russian Academy of Sciences since 1972 and has served as professor in applied mathematics at Moscow State University and various research institutes and universities abroad. He is well known for his visualization-based, multi-objective methods and integrated assessment modeling of environmental and water resources management problems including those of the River Volga and Lake Baikal. <avlotov@gmail.com>

Daniel (Pete) Loucks obtained degrees in forestry at Penn State and Yale Universities and then, some years later, studied environmental systems engineering at Cornell University. Cornell has always felt he needed to learn more so they have kept him, except for many stints in other jobs and places in the world to help broaden his experiences and perspectives and to have fun. He has gained much from his students and others he has worked with, including many of the authors of this book. <loucks@cornell.edu>

Ronnie S. Mckenzie has degrees in water resources planning from the University of Strathclyde in Glasgow. Dr. McKenzie has worked for more than 35 years in the fields of water demand management, hydrology, water resource planning, and management and operations in many countries, including South Africa, the United States, Ethiopia, New Zealand, Australia, and the United Kingdom. He also has advised many organizations, including the World Bank, the EU, and the UN. He is a fellow of the

International Water Association, the Water Institute of Southern Africa, and the South African Institute of Civil Engineering. The South African Government awarded him the first Ministerial award for supporting water loss reduction practices throughout South Africa. His contributions to developing countries won him two awards, one from the British government and the other from the International Water Association. His recent books include one on water losses, which took two years to write and sold 20 copies, and another on meteorites that took three weeks and sold more than 5,000 copies. Go figure! <ronniem@wrp.co.za>

Richard M. Males, Ph.D., went into civil engineering and water resources because he never wanted to work in an office, and he dreamed of building tall dams in exotic locations. Instead, he has spent all his professional life in front of an 026 keypunch, an ASR-33 terminal, or some kind of computer screen. He met Walter Grayman in 1967, when both were in graduate school at MIT, and they have been close friends and colleagues ever since. In 1985 they created the first ASCE Specialty Conference on computer applications in water resources. <males@iac.net>

H. P. Nachtnebel, O.Univ.Prof. Dipl.-Ing. Dr.Techn., is emeritus professor and head of the Institute of Water Management, Hydrology and Hydraulic Engineering, University of Natural Resources and Life Sciences, Vienna. He has taught, conducted research, and consulted internationally on topics, such as eco-hydrological modelling, river management and restoration, stochastic hydrology, water resources systems and risk analysis, climatology, groundwater management, and hydraulics and hydrology. <hans_peter.nachtnebel@boku.ac.at>

Lindell Ormsbee, Ph.D., P.E., P.H., is the Raymond-Blthye professor of civil engineering at the University of Kentucky. He directs the Kentucky Water Resources Research Institute and is the associate director of the university Superfund Research Center. His research has focused on the application of systems analysis methods to complex problems in water distribution and stormwater management systems, remediation of contaminated groundwater, disinfection technologies for wet weather discharges, stakeholder engagement and watershed management, and nutritional and behavioral strategies for reducing the impact of exposure to Superfund chemicals. <lindell.ormsbee@uky.edu>

Geoff Pegram, Ph.D., is emeritus professor in civil and environmental engineering at the University of KwaZulu-Natal, South Africa. He is the director of Pegram & Associates which specializes in hydrometeorological research. After earning his bachelor's of science in Engineering, he worked with Rhodesia Railways, followed by a year working on the South African Orange-Fish tunnel (one of the longest on Earth). He then joined his alma mater and has been in academia ever since, collaborating and publishing with like-minded people around the world. <Pegram@ukzn.ac.za>

Jerome Delli Priscoli chairs the Global Water Partnership (GWP) Technical Committee. For more than 40 years, he has been senior advisor at the US Army Corps of Engineers' Institute for Water Resources. He is on the board of governors and is a former member of the executive bureau of the World Water Council. He has been an advisor to the World Bank and UN water related agencies, as well as to foreign water ministries on water policy issues. He is also the editor-in-chief of the peer reviewed journal *Water Polic*y. He continues to facilitate dialog among diplomats, senior political leaders, and NGOs in the World Water forums. He holds degrees in economics and political science and post-doctoral studies in theological studies from Tufts and Georgetown Universities. <jdpriscoli@gmail.com>

David Purkey, Ph.D., is the Latin America regional director for the Stockholm Environment Institute based in Bogota. Prior to moving to Colombia, David spent 12 years leading the SEI–US Water Group, which has worked for more than 25 years to develop water resource modeling tools that can support water managers and stakeholders in the evaluation of water management options in specific river basins. Leading the effort to develop and deploy SEI's signature water software has afforded David the chance to work in river basins around the world, where he has discovered that water management decisionmaking is a fascinating balance between rigorous analysis and stakeholder engagement. <david.purkey@sei.org>

David E. Rosenberg, Ph.D., is an associate professor in the Department of Civil and Environmental Engineering at Utah State University. His research includes water conservation, managing water to enhance ecosystems, hydro-economics, hydroinformatics, and other topics that bring people and water together. When he's not working, he likes to recreate on water in its liquid and solid forms. <david.rosenberg@usu.edu>

Laurel Saito, Ph.D., P.E., is the Nevada Water Program director for the Nevada chapter of The Nature Conservancy. Prior to this position, she was an associate professor of water resources at the University of Nevada, Reno, and she has worked in the public and private sectors as a water resources engineer. She is a strong advocate of interdisciplinary approaches to water resources management. <laurel.saito@tnc.org>

Hubert Savenije is professor of hydrology at Delft University of Technology and was previously professor of water resources management at IHE-Delft. He started his career in 1978 as a hydrologist in Mozambique and subsequently joined an international consultancy firm as a water resources engineer. After having worked in different parts of the world, he completed his Ph.D. in 1992 and subsequently became an academic. His research and education have always remained closely connected to water resources issues in the developing world. <H.H.G.Savenije@tudelft.nl>

Andreas H. Schumann is professor for hydrology and water management at the Ruhr-University Bochum. He has worked eight years in several state water authorities in East Germany, but, at the beginning of 1989, he started a new scientific career in the western part of Germany. In academia, he never lost the contact to practice and still tries to find scientific based solutions for practical problems. <Andreas.Schumann@hydrology.ruhr-uni-bochum.de>

Uri Shamir is professor emeritus at the Technion, Israel. He returned to Israel in 1967 with a Ph.D. from MIT, and he had challenging and enjoyable times teaching and mentoring on hydrology and water resources systems at the Technion and during sabbaticals abroad. He met Chuck Howard while at MIT, and they became lifelong friends, generating ideas, implementing them in projects, and publishing papers together. He served as a consultant to the Israeli national water company Mekorot and to the Israeli Water Authority. He is also past president of two international professional organizations (IAHS and IUGG) and served as a member of the Israeli negotiating team on water with Jordan and the Palestinians. <shamir@technion.ac.il>

Pete Shanahan is a hydrologist and environmental engineer who has mixed consulting practice, university teaching, and research during his 40-year career. He is a retired senior lecturer from MIT and continues to consult part-time with HydroAnalysis, Inc., a firm he founded 30 years ago. <peteshan@mit.edu>

Oskar Sigvaldason, Ph.D., P.Eng., worked with Acres International, consulting engineers for 38 years, including nine years as president. He became a strong advocate of systems methodology after graduating as a civil engineer and completing a year of post-doctoral studies at Harvard University. His applications have been used in water resources systems, preparation of national energy and electricity supply plans for various countries, and most recently, defining national plans for mitigation of greenhouse gasses and associated transformations of energy systems. <oskar@sigvaldason.com>

Slobodan P. Simonovic is recognized for his unique interdisciplinary research in systems analysis and the development of deterministic and stochastic simulation, optimization, and multicriteria models for addressing the challenging system of systems problems lying at the confluence of society, technology, and the environment. Educated in Serbia (University of Belgrade) and in the United States (University of California), he teaches and is engaged in research and consulting at the University of Western Ontario (London, Ontario, Canada). <simonovic@uwo.ca>

László Somlyódy is professor emeritus at the Budapest University of Technology and Economics. Earlier, he was a department head at the same university. He has also served as head of the Hungarian Water Research Center, Vituki, and as a scholar in the water resources programme of IIASA (International Institute for Applied Systems Analyses). He is Hungary's "water quality man." The main corridor of Vituki's prestigious headquarters building on the Danube Embankment is now called the Somlyody Corridor. <somlyody@vkkt.bme.hu>

Gene Stakhiv was educated on the mean streets of the Bowery in New York City, and he later received his Ph.D. in water systems engineering at Johns Hopkins University. He spent his entire career with the US Army Corps of Engineers—most of it in research and policy studies at the US Army Institute for Water Resources, where, for the past 20 years, he was actively engaged in spreading the doctrine of comprehensive planning and integrated water resources management, both nationally and internationally. <Eugene.Z.Stakhiv@usace.army.mil>Mike

Jery Stedinger, Ph.D., NAE, Dist.M.ASCE, started out in applied math at the University of California, Berkeley, for a bachelor's of science, then went to graduate school at Harvard University, where Mike Fiering and Harold Thomas (and later Pete Loucks) introduced him to the challenges of water resource systems engineering and stochastic hydrology. He pursued those interests for 40 years as a professor at Cornell University with two sabbaticals, each with the US Geological Survey (USGS National Center) and the US Army Corps of Engineering (IWR and HEC). He believes that the science of water management moves forward fastest when people work together. <jrs5@cornell.edu>

Cris Surbeck, Ph.D., P.E., is the associate dean for academic and student affairs of the School of Engineering at the University of Mississippi.

Previously, she worked in environmental consulting in Southern California. Currently, she gets involved in all kinds of outreach programs about water, the environment, infrastructure, and higher education. <csurbeck@olemiss.edu>

Desiree Tullos, Ph.D., P.E., is a professor in the Biological and Ecological Engineering Department at Oregon State University, where her work emphasizes sustainable management of rivers and training the next generation of systems-oriented engineers. She is grateful for a career that provides excuses to explore and share new rivers, landscapes, relationships, and disciplines. <Desiree.Tullos@oregonstate.edu>

Alicia Wang is studying computer science at Cornell University and developing all kinds of useful applications. Her specialty is backend engineering, that is, creating software that make interfaces work, such as web pages that have to access data. One of her numerous other talents is drawing cartoons for Cornell University's daily newspaper. <axw5@cornell.edu>